Science in the New Russia

LOREN GRAHAM
AND IRINA DEZHINA

Science in the New Russia

Crisis, Aid, Reform

INDIANA UNIVERSITY PRESS
Bloomington and Indianapolis

This book is a publication of

Indiana University Press
601 North Morton Street
Bloomington, IN 47404-3797 USA

http://iupress.indiana.edu

Telephone orders   800-842-6796
Fax orders   812-855-7931
Orders by e-mail   iuporder@indiana.edu

The paper used in this publication meets the minimum requirements of American National Standard for Information Sciences—Permanence of Paper for Printed Library Materials, ANSI Z39.48-1984.

Manufactured in the United States of America

Library of Congress Cataloging-in-Publication Data

Graham, Loren R.
  Science in the new Russia : crisis, aid, reform / Loren Graham and Irina Dezhina.
     p. cm.
  Includes bibliographical references and index.
  ISBN-13: 978-0-253-35155-5 (cloth : alk. paper)
  ISBN-13: 978-0-253-21988-6 (pbk. : alk. paper) 1. Science—Russia (Federation)—History. 2. Science—Soviet Union—History. I. Dezhina, Irina, date- II. Title.
  Q127.R8G73 2008
  509.47′09049—dc22

                                                                                   2007047371

1 2 3 4 5   13 12 11 10 09 08

Contents

Introduction

What happens to science when the nation supporting it disappears? What is the relationship between the political structure of a state and the way it conducts science and higher education? What are the effects—intended and unintended—of massive foreign assistance to a nation's science?

These questions are at the center of this book. All emerged as pressing issues after the collapse of the Soviet Union, when one of the world's largest and most impressive science establishments was suddenly cut adrift. The slashing of the federal science budget to approximately 20 percent of what it had been in the Soviet period, combined with the departure of thousands of the most talented scientists, damaged Russian science so seriously that some observers predicted its demise.[1] We now know that this prediction was incorrect; Russian science has survived and to some degree is recovering.

Recovery, however, is not the same as restoration of what was before. The Soviet centralized and autocratic system of organizing and financing research was inappropriate for a democratic, free market society. (Russia's inability to fully become such a society is one reason the debates about science remain so controversial even today.) Russia's effort to find new ways to organize and support science that are congruent with its new political and economic order has been plagued with difficulties. But, then, just what is the relationship between science and democracy? The best science is unapologetically elitist. And the autocratic Soviet Union probably did better in science than in any other of its major activities. The question of how much the governance and financing of science should change in the New Russia has been much more perplexing than, say, the question of how much the centralized state-controlled economy should change. The virtues of moving from a centrally controlled economy to a market one are obvious throughout the world. The virtues of moving from an elitist science system to a more decentralized one have been seen much less clearly in Russia, and indeed, perhaps elsewhere as well.

In the last twenty years both the financial supports and the organizational principles of Russian science have been fundamentally questioned. This book describes and analyzes these challenges and the changes that have been made in response to them. It also places the Russian experience in the context of world trends in the organization of scientific research, comparing the ways science has been structured and governed in other leading nations with the situation in Russia. We observe that Russia, in the organization of its science, has been and still is out of step with many of the countries competing with it scientifically, technologically, and economically.

One important influence on these recent events in Russian science has been foreign foundations and organizations, which have been engaged in what one U.S. foundation official called "perhaps the largest program of scientific assistance the world has ever seen."[2] Foundations and organizations headquartered in more than twenty different countries have pumped the equivalent of several billion U.S. dollars into Russian science. But these foundations and agencies did not just give money; they brought their own political, even ideological, principles, and these have been important factors in the debates about change within Russian science and higher education. In the process of rendering aid, foreign foundations have introduced principles that were entirely new to Russia, such as peer review and open competition for grants. Some have consciously tried to turn Russian science away from military purposes. Others have not only supported science, but also fostered democracy, human rights, and civil society in Russia as a whole. Because of these political agendas, the Putin government, which came to power in 2000, has been suspicious of the foreign foundations and has recently passed legislation designed to exercise control over their activities. Of course, not all foreign foundations and organizations have pushed for political change; some are primarily interested in mutually beneficial cooperation between the technical and scientific communities of Russia and other countries, usually those in which the foundations and organizations are based.

We would like to emphasize that this book focuses on research in the natural sciences, not on the social sciences, higher education, international aid in general, brain drain, or technology policy (all of which have indirect connections to our theme and which therefore are briefly discussed).

We outline, first, the basic organization of science in the Soviet period, the backdrop for all that followed; then we describe and analyze the immediate post-Soviet crisis and the attempts to deal with it on the part of both the Russian government and foreign foundations and agencies that rendered assistance. These attempts still continue today, and a final assessment of their results can not yet be written. Nonetheless, patterns have emerged that are not likely to change in the short term. In our concluding chapter we describe these patterns and give our overall evaluation of what has occurred in Russian science in the last twenty years. In this introduction, we will give only a taste of those conclusions and urge readers to turn to the final chapter for a fuller elaboration.

Russian science has survived and has found a new footing. Science budgets are now increasing and brain drain has lessened, even though it remains a problem. Further reforms in the management, organization, and funding of Russian science are still needed, particularly in Russia's effort to create a commercial culture in which innovations are quickly utilized in the economy. This reform has barely begun, although in late 2006 for the first time substantial funds were designated for programs with that goal.

Another difficult reform for Russian science is uniting teaching and research. The prestige of, and favoritism toward, the Russian Academy of Sciences, a vast research organization, has made the creation of "research universities" in Russia particularly difficult. Foreign, and especially U.S., help has been significant in the effort to strengthen Russian universities as both teaching and research centers.

Although there has been much news recently indicating that the Academy of Sciences has been "subordinated" to the government, in actual fact significant changes to the Academy have not yet been made. The recent legislation giving the confirmation of the president of the Academy of Sciences to the president of the Russian federation (Putin and his successors) is a return, *de facto,* to practices in the Soviet period. By itself it does not change very much, although it is further evidence of the centralizing tendency of the Putin government. The Academy has proved resilient in dominating basic research, even if it does not have political independence.

Western foundations, especially those in the United States but also foundations and organizations in Western Europe, North America, and Japan, have had a major impact on Russian science. They helped save Russian science from collapse in the mid-nineties. The largest lasting impacts of these foundations have been the introduction of peer review and the funding of research with competitive grants, practices that were earlier unknown in Russian science.

The very fact that this book is co-authored by an American, based in Cambridge, Massachusetts, and a Russian, based in Moscow, is an illustration of the dramatic effect on scholarship of the end of the Cold War. Twenty years ago such a book containing both scientific and political elements could not, for political reasons, have been written jointly by an American and a Russian. But not only has politics changed; without the technological innovation of the internet and electronic mail we simply could not have engaged in the daily discussions and exchanges of drafts necessary for writing such a book.

* * *

We are grateful for the financial assistance of the National Science Foundation (Award 0217597) for support of our research. Although we have not received research grants from them for this book, the John D. and Catherine T. MacArthur Foundation, the Civilian Research and Development Foundation, the Carnegie Corporation of New York, and a number of other foundations and organizations have been very helpful to us. In the early nineties the Sloan Foundation gave a grant to Loren Graham to collect materials on events in Russian science that have been valuable to us.

In conducting research for this book we have had dozens of interviews, site visits, and contacts with scientists and science administrators, mostly in Russia, but also in Western Europe and the United States. We have interviewed officials from the Ministry of Education and Science, the Russian Academy of Sciences, and more than twenty universities in Russia, not only those in the "academic centers" of Russia in Moscow, St. Petersburg, and Novosibirsk, but also in several dozen other cities, such as Voronezh, Saratov, Kazan, Rostov, Ekaterinburg, Krasnodar, Perm, Tomsk, Vladivostok, Nizhnyi Novgorod, Samara, and elsewhere. We have visited and talked to administrators and scientists at the Russian Foundation for Basic Research, the International Science and Technology Center in Moscow, the Alexander von Humboldt Foundation, the Moscow office of the MacArthur Foundation, the U.S. National Institutes of Health, the German Academic Exchange Service, the Deutsche Forschungsgemeinschaft, the Max Planck Society, the Royal Society in London, the International Research and Exchanges Board, the National Science Foundation, George Soros, the Information-Science-Education Center in Moscow, the Netherlands Organization for Scientific Research, the Organization for Economic Cooperation and Development in Paris, the International Association for the Promotion of Cooperation with Scientists from the Independent States of the Former Soviet Union, the European Community, Pro-Mathematica in France, and elsewhere.

We have drawn upon the valuable research done by others on specific topics related to this research. Particularly helpful have been the works of Harley Balzer, Glenn Schweitzer, and Gerson Sher.[3] In addition to these individuals, we are indebted to dozens, even hundreds, of others for conversations and interviews—far too many to name. We would like to thank, in particular, Victor Rabinowitch, Marilyn Pifer, Marjorie Senechal, John Slocum, Deana Arsenian, Boris Saltykov, and Mikhail Strikhanov. Specific works and conversations will be referenced in the text.

The co-authors would like to thank their home institutions for providing support, material and intellectual, for their research; in the case of Loren Graham, these are the Program on Science, Technology, and Society of the Massachusetts Institute of Technology, and the Davis Center for Russian and Eurasian Studies of Harvard University; for Irina Dezhina it is the Institute for the Economy in Transition in Moscow. Helen Repina at the Davis Center Library was particularly helpful.

Throughout the book we have, through necessity, used a number of acronyms (e.g., RFBR for Russian Foundation for Basic Research). The first time we use such an acronym we try to explain its full meaning, but at later moments we often use only the short form. On pp. xi–xiii there is a list of acronyms and explanations to which the reader can refer as needed.

Acronyms

AID	U.S. Agency for International Development
ANVAR	Agence Nationale de Valorisation de la Recherche; French state agency for innovation
AUTM	Association of University Technology Managers, headquartered in Northbrook, Illinois
BRHE	Basic Research and Higher Education program
CASE	Centers of Advanced Study and Education
CCU	Centers for the Collective Use of Equipment
CERN	Conseil Européen pour la Recherche Nucléaire, European Organization for Nuclear Research
CGP	Cooperative Grants Program
CIS	Commonwealth of Independent States
CNRS	Centre National de la Recherche Scientifique
CRDF	U.S. Civilian Research and Development Foundation
CSRS	Center for Science Research and Statistics of the Russian Ministry of Science and Education and the Russian Academy of Sciences, Moscow
DAAD	Deutscher Akademischer Austausch Dienst, German Academic Exchange Service
DESY	Deutsches Elektronen-Synchroton
DFG	Deutsche Forschungsgemeinschaft, German Research Foundation
FKTsB	Federal Commission for Securities
FSA	Freedom for Russia and Emerging Eurasian Democracies, and Open Market Support Act of 1992, U.S. Congress
FSU	Former Soviet Union
FTsNTP	Federal'naia Tselevaia Nauchno-Tekhnicheskaia Programma, Federal Goal-Oriented Program in Science and Technology
FZ	Federal'nyi Zakon, Federal Law
GK	Grazhdanskii Kodeks, Civil Code
GNTs	Gosudarstvennyi nauchnyi tsentr, Federal Research Center
GDP	Gross domestic product
IET	Institute for the Economy in Transition
INO	Informatsiia-Nauka-Obrazovanie, Information-Science-Education Center
INPRAN	Institute of Economic Forecasting, Russian Academy of Sciences

INTAS	International Association for the Promotion of Cooperation with Scientists from the Independent States of the Former Soviet Union
INTERROS	Large Russian private investment company
IP	Intellectual property
IPR	Intellectual property rights
IREX	International Research and Exchanges Board
ISF	International Science Foundation
ISSEP	International Soros Science Education Program
ISTC	International Science and Technology Center
ITC	Innovation Technology Center
MEI	Moskovskii Energeticheskii Institut, Moscow State Power Institute
MIET	Moscow Institute of Electronic Technology
Minpromnauki	Ministry of Industry, Science, and Technology
NGO	Non-governmental organization
NIH	U.S National Institutes of Health
NIS	National Innovation System
NSF	U.S. National Science Foundation
NWO	Nederlandse Organisatie voor Wetenschappelijk Onderzoek, Netherlands Organization for Scientific Research
OECD	Organization for Economic Cooperation and Development
PI	Principal investigator
PPP	Public-private partnership
PDST	Priority Directions of Science and Technology
R&D	Research and development
RAN	Rossiiskaia Akademiia Nauk, Russian Academy of Sciences
REC	Research and Education Center
RESC	Regional Experimental Science Center
RF	Russian Federation
RFBR	Russian Foundation for Basic Research
RFH	Russian Foundation for the Humanities
RFTD	Russian Fund for Technological Development
RUDN	Rossiiskii Universitet Druzhby Narodov, Russian University of People's Friendship
RVC	Russian Venture Company
SBIR	Small Business Innovation Research, U.S. program
SEZ	Special Economic Zone
Sibneft	Large Russian oil company
SWOT	Strengths, weaknesses, opportunities, threats
TACIS	Technical assistance program of the European Union
TCS	Teaching Company Scheme
TsISN	Tsentr Issledovaniia i Statiski Nauki, also CSRS (Center for Science Research and Statistics)
TTO	Technology Transfer Office

VIF	Venture Innovation Fund
Vympelkom	Russian communications company, Bee-Line Network
YUKOS	oil company, named for: Yuganskneftegaz-Nefteyugansk-KuibishevnefteOrgSintez

Science in the New Russia

1 Science at the End of
the Soviet Period

To the eye of the foreign observer, two of the most notable characteristics of the science establishment of the former Soviet Union were its bigness and its high degree of centralization. Not only was the Soviet science and technology community one of the largest in the world, but it was organized in distinctly different ways from those in other countries. The relative roles of the academies of sciences, the universities, and the industrial research organizations were unusual when compared with those in other nations. Furthermore, Soviet science and technology were organized in larger units than found elsewhere, and under the control of fewer influential individuals.

At the beginning of 1991, the year in which the Soviet Union collapsed, the total number of people officially classified in the Soviet Union as scientific researchers and faculty was 1,520,000.[1] Because of definitional problems, this number cannot be directly compared to similar statistics in other countries, but there is general agreement among Western specialists that by the 1980s the Soviet Union had 10 to 30 percent more scientists and engineers than the United States, depending on the definition of degrees and fields.[2] When one considers that the Soviet economy at its best moment was much smaller (estimates say less than one-half the size) than the U.S. economy, one is tempted to conclude that the Soviet Union had too many scientists and engineers, but such a conclusion would be of course disputable since no one knows what the optimal number of scientists and engineers is for a given economy. But certainly the financial burden on the Soviet Union to maintain this enormous scientific establishment was very great.

These Soviet researchers were distributed in three gigantic pyramids, which for sake of convenience can be called the university system, the academy of sciences system, and the industrial and defense ministry system. Table 1.1 gives a very approximate description of the organization and shares of research personnel and budgetary funds of each of the three pyramids.[3]

From the standpoint of U.S. experience, the most familiar pyramid of the three is the university system. The Soviet Union, like the United States, had large universities, all supported by the government in the Soviet Union, as the majority are partially supported by governments (usually state governments) in the United States. (Of course, the United States also has a number of quite influential private universities; there were no private universities in the Soviet Union.) The superficial similarity of the two government university sys-

Table 1.1. Organization of research personnel in the Soviet Union, ca. 1990

University system	Academy of Sciences system	Industrial and defense system
State Committee of Higher and Secondary Education of USSR	Academy of Sciences of USSR	Industrial ministries; defense ministry
Union-republic ministries; committees of higher and secondary education	Siberian Division of the Academy in Novosibirsk; other branches and filials	
Higher educational institutions (vysshie uchebnye zavedeniia): universities and colleges, including leading universities (Moscow, Leningrad): 770	Academies of sciences of union republics: 14 Academies of Agricultural Sciences; Medical Sciences; Pedagogical Sciences; Engineering (new)	Industrial research institutes (otraslevye instituty**); closed military ("postbox") research institutes
600,000 researchers*	125,000	800,000
7% of R&D budget	6.5% of R&D budget	87% of R&D budget

* Includes faculty members as well as researchers.
** E.g., institutes of steam turbines, coal mining, electronics.
Source: D. Piskunov, "Soviet Fundamental Science: State, Problems, and Perspectives of Development" (unpublished manuscript, Analytical Center for Problems of Socio-Economy and Science-Technology Development, Academy of Sciences of the USSR, Moscow, 1991.)

tems should not blind one to the enormous differences. In the United States the universities are the home of most of the fundamental research conducted in the country. In the Soviet Union the universities traditionally had a much narrower pedagogical role, and the research conducted in them was often of an applied character.[4] Another difference is that even state universities in the United States receive large portions of their budgets from non-governmental sources.

The Academy of Sciences of the USSR was the most unusual institution, from the perspective of the United States. The Soviet Academy of Sciences was the home of fundamental research and the most prestigious scientific institution in the country. In the United States the National Academy of Sciences and the National Academy of Engineering are primarily honorific organizations; they issue reports periodically, but they are neither the locus of laboratory research nor the place of employment of active researchers. The Soviet Academy of Sciences, on the other hand, was the place of employment of the most outstanding fundamental researchers in the country. These scientists spent their lives in its service, and their places of residence, travel and vacation privileges, and health and social services were traditionally controlled by it. No surprise that Alexander Vucinich called it an "empire of knowledge."[5]

The industrial and defense ministerial system was by far the largest of the pyramids. It was primarily concerned with applied science, although it performed some fundamental work as well (just as the Academy of Sciences system performed some applied work). Military research occupied a very large role, not only in this industrial-defense pyramid, but in the universities and academy institutes as well. In fact, the military was given about 75 percent of all resources.[6] Thus, when we say that the Academy of Sciences was the home of fundamental research we are making a relative statement. Even the Academy was heavily influenced by military and applied demands.

Most researchers in the industrial-defense pyramid were not located in individual plants or "companies" (as they would be in the United States), but instead, like their colleagues in the academies of sciences, they usually worked in centralized institutes in large cities. In the period of the most intense growth of the Soviet scientific establishment (1960–1972) the number of these institutes increased from 4,196 to 5,307.[7] After 1972 the number of institutes remained approximately stable, but the existing institutes continued to grow in size. The average institute in 1991 had a staff of 270 researchers, but some of the most important had research staffs of several thousands. The directors of these institutes were traditionally very powerful, even autocratic, figures appointed from above, although in the last years of the Soviet Union, under Gorbachev, an attempt was made to introduce democratic elections of directors.

In the above paragraphs we have several times compared the Soviet system of the organization of science with that of the United States. We have made these remarks on the assumption that many readers of this book will be Americans for whom such a comparison is informative. It would be perfectly legitimate to propose, however, that the most appropriate model of comparison for the Soviet system would be that of one of the continental European countries, such as France or Germany, where the role of the government in administering and financing the science establishment is very large, as was the case in the Soviet Union and is the case in Russia today. Later on in this book we will raise this issue in more detail and make brief comparisons of Russia to these other countries. At this point, however, we will merely note that many Russian science policy analysts today are aware of these different models of comparison. On the one hand, they often note that Russia's science system is more similar to those of several European countries than it is to the United States; on the other hand, they believe that U.S. experience is especially valuable for Russia. The United States and the USSR were, after all, the two Cold War superpower rivals in science and technology, and they have been heavily dependent on defense-related R&D to build up their capabilities in science. Therefore, both countries face, to a certain extent, similar problems, such as knowledge transfer between the government and public sectors, and linking military research more closely with civilian needs. Furthermore, Russian science policy analysts know that France and Germany are also trying to copy some models and practices from the U.S. science establishment, demonstrating interests similar to their own.

How Good Was Soviet Science?

The Soviet scientific system possessed both strong points and weak points. Among the strong points were generous governmental and social support for the natural sciences, both financial and psychological; a strong educational establishment, many of whose most academically talented graduates went into the research institutes; and the ability, through political control and the command economy, to concentrate successfully on a few high-priority projects (e.g., nuclear weapons, space). Among the disadvantages of the Soviet scientific system were the separation of research and teaching (a result of the division of functions in the three pyramids already described); the distortion of priorities, particularly toward the military; the low productivity of the research system (especially when one considers the enormous resources devoted to science), a flaw that was connected to the absence of genuine peer review; political restrictions (secrecy, repression of dissidents, prejudice against some ethnic groups, such as Jews, and suppression of certain fields, such as genetics from 1948 to 1965); the pervasiveness of corruption,[8] and, finally, an emphasis on "reverse engineering" of Western innovations. (The reverse engineering approach was not always a disadvantage, since Soviet engineers proved adept at first copying and then sometimes improving on Western technology).

In some fields Soviet science was outstanding. In 1977 the National Academy of Sciences in the United States conducted a thorough study of the quality of Soviet science.[9] Most of the members of the evaluating committee were distinguished U.S. natural scientists who were familiar with Soviet work in their fields.

The striking aspect of the evaluations given by U.S. scientists of Soviet work in a number of fields was the excellence they found there. Lipman Bers, past president of the American Mathematical Society, professor of mathematics at Columbia University, and a native reader of the Russian language, described Soviet mathematics in the 1970s as equal to mathematics in any other country of the world, and said that Moscow probably contained more great mathematicians than any other city on the globe. Bers was fully aware of past and continuing persecution of scientists in the Soviet Union (this was the time of refuseniks in Soviet mathematics, the firing of mathematicians who applied to emigrate to Israel). But, he said, the strength of Russian mathematics was so great and the pool of talent so deep that repression could not destroy it.

In a similar mood, David Pines, professor of physics at the University of Illinois, reported that "in the area of theoretical condensed matter physics, USSR scientists have been doing some of the most innovative and important work in the world for two decades or more." Hans Frauenfelder, another physicist from Illinois, wrote that "the level of the best work in solid state theory in the USSR is outstanding and at or near the frontier."

Similarly strong comments were made about other fields in the field-by-field reports submitted by outstanding U.S. scientists, most of them members of the National Academy of Sciences. Fields of Soviet science selected for particular praise by U.S. scientists in the late 1970s included, in addition to mathematics, plasma physics, theoretical seismology, climate research, and theoretical astrophysics.

Soviet science was obviously very good in some fields. However, if one takes into account the enormous financial effort being made by the Soviet Union to excel in science, the conclusion seems inescapable that it was underperforming. György Péteri of the Norwegian University of Science and Technology has attempted to measure "how hard countries were trying" in their research efforts by plotting the ratio of the density of research staff (full-time equivalent number of research workers per 10,000 inhabitants) against the level of economic development (GDP per capita).[10] He made these calculations for eighteen different countries, both capitalist and socialist, for the year 1985. According to his analysis, two countries—East Germany (the German Democratic Republic) and the Soviet Union—were the most extreme examples of what he called "systemic overstretch." These two countries were trying extraordinarily hard in science and investing a great deal of capital, both personal and financial, in research.

The mature Soviet Union possessed a very large scientific establishment, by some calculations the largest in the world.[11] However, no matter what criteria of excellence one chooses—number of Nobel Prizes awarded, frequency of citation of Soviet research, number of inventions registered abroad, or honorary memberships in foreign scientific societies—the achievements of Soviet scientists were disproportionately small.[12] In the fields of natural science (physics, chemistry, physiology, and medicine) from 1901 through 1990, citizens of the United States won 145 Nobel Prizes, compared to 8 received by citizens of Russia and the USSR (after the collapse of the USSR, in the year 2000, Zhores Alferov won a prize in physics for work done in the Soviet period, and Vitaly Ginzburg in 2002 also won such a prize, making the total 10). If one ranks the countries of the world in terms of the number of their citizens who received Nobel awards in this period, the Soviet Union ranked sixth, after the United States, the UK, France, Germany, and Sweden. And the Soviet Union seemed to slip badly in its last decades. Between 1976, when Peter Kapitsa received a Nobel award, and 1991, when the Soviet Union collapsed, there were no Nobel Prizes given to Soviet natural scientists.

If one looks at the sale of licenses for technology, the poor performance of Soviet research is particularly clear. The United States, in sharp competition with Western Europe, Japan, and other advanced industrial countries, sold approximately thirty times as many licenses for technology annually as the Soviet Union in the last years of its existence. A similar picture emerges from a study of honorary memberships in prestigious scientific societies. In 1986, five years before the disappearance of the Soviet Union, the Royal Society in London had 87 foreign members, of whom 6 were citizens of the Soviet Union and 44 citizens of the United States. During the next two years, 3 of the Soviet members died, leaving only 3. The situation for the Soviet Union was slightly better among foreign members of the National Academy of Sciences in the United States, in which 16 members were Soviet citizens in 1986, out of a total of approximately 250 foreign associates.

Each criterion for comparing science in the Soviet Union with that in other countries has, of course, its flaws. The Russian language was known by few researchers outside the Soviet Union, and consequently the achievements of Soviet researchers were more frequently overlooked than those presented in more accessible languages. Furthermore, comparisons of citation rates of scientific publications in the United States and the USSR are inaccurate because so much of Soviet science was secret and therefore could not be cited. Also, it is hardly surprising that so many more U.S. scientists were members of the Royal Society than Soviet scientists; the common language and many ties shared by British and U.S. scientists make such shared honors inevitable.

Nonetheless, no matter what criteria of excellence one chooses, Soviet science failed to measure up to that of several other countries—and that fact has significance. After all, in the first decades of Soviet history the view was fairly widespread, both in Soviet Russia and among left-wing scientists in Western countries, that Soviet science would become the best in the world. The Soviet leader Nikolai Bukharin wrote in 1931, "It is not only a new economic system which has been born. A new culture has been born. A new science has been born."[13] Soviet sympathizers in the West such as J. D. Bernal and J. G. Crowther wrote books praising Soviet science as superior in essence to that in the West.[14] Bernal wrote in 1939 that Soviet technology and science would serve humanity much better than that of "the present indefensible and chaotic system of science and industry in the West." As one observer commented, Bernal was "inclined to accept the claim of Soviet Marxism to represent science in general."[15] The distinguished U.S. historian of medicine Henry Sigerist was so impressed with Soviet medical science in 1937 that he wrote that it was the "beginning of a new period in the history of medicine."[16] Even as late as 1968 the British historian of science Joseph Needham remarked that "all those in the thirties who believed that the natural sciences could only come to their most perfect fruition in a socialist society probably still think so now."[17] Against the background of such hopes the conclusion is inescapable that Soviet science and technology did not meet the grand goals its early supporters posed. Indeed, such statements about Soviet science as those quoted in this paragraph now seem naïve and grossly uninformed.

Rather than being the best science in the world, science in the late Soviet Union was crying out for reform. It was a system that emphasized quantity over quality, seniority over creativity, military security over domestic welfare, and orthodoxy over freedom.

World Trends in the Organization of Science

The question that has absorbed the minds of many scientists and science administrators in Russia in the last twenty years—how *should* scientific research be organized?—cannot be understood outside of world trends. Although no one knows with certainty the answer to the question of what is the best way to organize research, trends in the last three hundred years in countries leading in science point in certain directions. In the eighteenth century "academies" were often seen as the best loci for the work of a country's leading scientists. In the nineteenth century, starting in Germany and then elsewhere, research universities became more and more important and gradually pushed aside the academies as the actual place of research in most countries. In the twentieth century the idea began to grow that teaching in the universities constituted a "load" upon talented scientists that prevented them from devoting themselves properly to research, and therefore the concept of the non-teaching "research institute" gained prominence. In France, one of the early models of this new type of organization was the Pasteur Institute (1887); in Germany, an outstanding example was the Koch Institute (1891). In the United States emerging private foundations and wealthy individuals helped answer the call of leading scientists for new preserves for unfettered research. John D. Rockefeller backed the creation in 1901 of the Rockefeller Institute in which he said he wanted scientists "to work in utter freedom."[18] There would be no bothersome students, no teaching loads, few committee meetings, no outside obligations. Several other such institutes followed in the United States, including the Carnegie Institution of Washington (established one year after the Rockefeller Institute, in 1902), and quite a bit later, in 1930, the Institute for Advanced Study in Princeton. These institutions were modeled on the new European research institutes, particularly the German and French ones. In the case of the Rockefeller Institute, the Koch and Pasteur institutes were specifically cited by the original Rockefeller trustees as they created their new research organization.[19] In Germany, institutes as a model for research received an important new impetus with the creation in 1911 of a whole network of them in the Kaiser-Wilhelm Gesellschaft, the predecessor of the present-day Max-Planck Gesellschaft.

In the new Soviet Union of the 1920s the idea of planned research conducted by the government was especially prominent, and the new non-teaching research institutes seemed to be the way to do this. After returning in 1926 from an inspection tour of scientific institutions in Germany, France, and England, S. F. Ol'denburg, permanent secretary of the Academy of Sciences, wrote a report for the Soviet government in which he said, "If the eighteenth century was the century of academies, while the nineteenth was the century of universities, then the twentieth century is becoming the century of research institutes."[20] The new Soviet Academy of Sciences, containing at first dozens and then later hundreds of such institutes, was planned to be different from the learned societies of the eighteenth century, and also different from the universities of the nineteenth century; instead it would be a sort of "ministry of science" based on research institutes in which scientists did not have teaching duties but would instead devote themselves to the advancement of knowledge.

For a while it looked like the creation of such networks of non-teaching institutes was a worldwide trend. In France in 1936 a left-wing government came to power, dominated by socialists and communists. This government was very impressed by the new science establishment of the Soviet Union, and it created a scientific organization in France partly modeled on it that still exists today: the Centre National de la Recherche Scientifique (CNRS), which is a network of governmentally supported non-teaching institutes.

However, as time went on, the allure of the non-teaching institutes began to wane. The feeling began to spread among some academic administrators in the West that teaching, far from being an impediment to research, was actually a stimulus to it. Senior researchers who did not have to meet the challenges to their ideas that students sometimes bring often settled into intellectual ruts in which they pursued the same idea endlessly. Non-teaching institutes in the United States, such as the Rockefeller Institute in New York City and the Institute for Advanced Study in Princeton, may have done well, but they definitely did not fulfill the grand dreams of their originators to be the leading scientific institutions in the country.

In the United States the research universities did not decline as Soviet planners had predicted, but instead developed into the most powerful engines of scientific research the world has ever seen. The infusion of federal money that came during and after World War II played an enormous role here. Universities and the federal government in the United States developed a remarkably successful complementarity of functions in the promotion of science. The competitive peer-review process required in granting much of the federal money awarded to university researchers provided a mechanism for ensuring quality research. It is difficult for the director of an institute to diminish the budget of a veteran member of his or her research staff, even if that scientist is obviously not doing as well as previously; it is easy for the National Science Foundation to turn down the proposal of a university researcher if the peer-review panel submits critical assessments. The system of funding research forming in the United States after World War II was simply more effective than older and more traditional systems, both at home and abroad.

A symbolic indication of shifting attitudes toward the relative merits of non-teaching research institutes and research universities came in 1953 when the board of trustees decided to convert the non-teaching Rockefeller Institute in New York City into a university. The trustees at that time conducted a full examination of the achievements of the Rockefeller Institute and decided that it was not fulfilling its full potential. It is interesting to look at the comments that were made in the reports to the Rockefeller board at that time and compare them with the following recent criticisms that have been made of the research institutes of the Russian Academy of Sciences (RAN), which were (and are) primarily non-teaching research establishments:[21]

"Research groups in the Institute were 'cloistered' and 'isolated.'"
"Too much inbreeding of ideas and people."
"The atmosphere in the research labs does not have the freshness that the youthful enthusiasm of students can bring."

"Students have zany ideas, but out of every 100 such ideas, one or two turn out to be fundamentally important."

As a result of their reassessment the Rockefeller board decided to convert the Rockefeller Institute into Rockefeller University, complete with graduate students. Quite a few of the existing researchers disagreed with this decision and refused to take on students, but within five years it became clear that they were losing out in winning grants to their faculty colleagues who had students in their labs. This was a revelation, as the non-teaching senior researchers at first considered their teaching colleagues to be less distinguished intellectually. But these teachers were more successful in getting grants, they were creating more exciting laboratories, and they were publishing more valuable results. Once this difference became clear, the resisters capitulated, and the conversion of the Rockefeller Institute to Rockefeller University was complete.[22]

One more piece of evidence in favor of the thesis that universities excel over non-teaching research institutes can be found in the experience of another set of research institutes that exist in the United States: the National Institutes of Health (NIH). The NIH sponsors two types of research: intramural and extramural. Intramural is that research done in house, within one of the institutes of the NIH. Extramural research is that done with NIH money awarded outside in the form of competitive grants, usually to university researchers. It is very common to hear among health researchers in the United States (although admittedly this is a subjective observation) that the money given outside the NIH is much more productive than that awarded inside the organization. The controllers of the NIH budget seem to realize the validity of this observation when they dictate that the larger portion of NIH's funds (80 percent in 2005)[23] is devoted to extramural training and research, also agreeing with the desire of members of Congress to fund research in their own states and districts. Overall, the bureaucratic federal research institutes seem less nimble and less energetic than university and commercial laboratories.

In the United States the preference for research universities over non-teaching research institutes as the locus for both teaching and research has in recent decades become more and more clear. When Richard Feynman, a world-famous physicist, was offered a non-teaching position at the Institute for Advanced Study in Princeton, he turned it down. Feynman observed, "students are often the source of new research. . . . Teaching and students keep life going, and I would never accept any position in which somebody has invented a happy situation for me where I don't have to teach. Never."[24] Another example of this preference can be found in the views of Roald Hoffmann, Nobel-laureate professor of chemistry at Cornell University (and a foreign member of the Russian Academy of Sciences) who in 1996 wrote, on the subject of teaching and research, "not only are the two inseparable, but teaching makes for better research. . . . I am certain that I have become a better investigator, a better theoretical chemist, because I teach undergraduates."[25]

The Soviet Union's own history of science gave an example of the importance of combining teaching and research even on the lowest undergraduate level, but unfortunately this example was not heeded. Mathematics was probably the Soviet Union's strongest field, and the Moscow School of Mathematics its greatest scientific achievement. The origin of this school was a group of students at Moscow University, many of them undergraduates, who clustered around two professors—Dmitrii Egorov and Nikolai Luzin—in the early 1920s. This group is remembered today as the legendary "Lusitania."[26] When Lev Shnirel'man, who eventually made significant contributions to the calculus of variations, joined Lusitania he was just fifteen years old. Andrei Kolmogorov, one of the great mathematicians of the last century, was seventeen when he first came into the classes of Egorov and Luzin. Many other very young people joined Lusitania and built it into a movement that was of international significance. Lazar Liusternik, Pavel Uryson, Nina Bari, and Pavel Aleksandrov—all mathematicians of note—joined Lusitania at the ages of eighteen or under. Yet despite this brilliant illustration of the importance of combining teaching and research, the Soviet Union later tended to isolate some of its best mathematicians in research institutes of the Academy of Sciences, such as the Steklov Institute of Mathematics.

In France and Germany, where the non-teaching research institute as a model for organizing research was more powerful than in the United States, the feeling also has grown over the years that these institutes are not as productive as they should be, and do not compete adequately with the research university system in the United States.[27] In recent years, French and German administrators have admitted these problems, and there have been many efforts to bring the research institutes of the CNRS and the Max-Planck Gesellschaft closer to the universities, providing more opportunity for university faculty to have research time in the institutes, and asking the institute researchers to do some teaching in the universities. In Germany, there has recently been an attempt to create a system of "elite universities" that can more successfully compete with U.S. research universities.[28] But so long as the Max-Planck Gesellschaft exists in Germany and the CNRS exists in France it is very difficult for either country to create research universities of international rank. On this issue, France, Germany, and Russia share a problem: the historical legacies of the CNRS, the Max-Planck Gesellschaft, and the Russian Academy of Sciences stand in the way of their doing what historical evidence now suggests needs to be done: creating high-quality research universities where research and teaching are combined by talented professors and students.

Perceptive scientists and science administrators in the waning days of the Soviet Union were aware of these world trends, and the feeling began to grow that Russia had made a historic mistake in the way it organized scientific research. It had enthroned the idea of the non-teaching research institute more than any other leading scientific country in the world. It also went further in separating the important functions involved in the scientific and educational enterprise; teaching, fundamental research, and applied research were each centered in different organizational networks. Furthermore, it was the only leading nation in the world in which two functions in science were tightly combined: the awarding of the highest honors (full membership in the Academy of Sciences), and the actual administration of scientific research (directing the Academy institutes).

Gradually the degree to which Russia was out of step with world trends in the organization of scientific research became clear, at least to some Russian scientists. This perception is the background for the great debates over the organization of research that broke out as the Soviet Union itself began to disintegrate.

Before concluding this section we would like to note several ironies in the story we have just told. On the one hand, the U.S. system of combining research and teaching in universities seems to be a particularly productive system for producing knowledge. Post-Soviet science administrators in Russia now worry about how they can match this productivity. On the other hand, one of the reasons for the strength of U.S. science in the last two generations has been the increases in funding spurred by competition with successful Soviet nuclear, space, and military programs in the Cold War period. The two systems have definitely stimulated each other, in different ways, and in different times.

In addition, we note that the tendency to separate teaching and research in the Soviet system not only deprived researchers of the stimulus that students can bring (of the sort emphasized by the U.S. Nobel laureates Feynman and Hoffmann), but it also tended to depress the quality of Soviet universities, in which faculty members usually carried very heavy teaching loads and were neither encouraged nor expected to conduct research.

Both kinds of systems have their strengths and weaknesses. Some Soviet/Russian scientists argue that since Academy institutes *did* have graduate students (*aspiranty*), they were more efficient research and teaching institutions than U.S. universities in which faculty have to "waste" their time teaching undergraduates who will never themselves be scientists. Despite the rhetorical commitment to equality, the Soviet system of science and education was actually more elitist than that of the United States. This argument that the Soviet system was more efficient because it usually put prominent researchers together only with committed graduate students overlooks the fact that in at least some fields—mathematics and computer science are prominent examples—undergraduates are often extremely creative, stimulating the research of their professors in ways it might not have been otherwise.

In sum, these questions of how research and teaching should be related to each other are complicated, with no easy answers. The historical evidence in the early twenty-first century seems to show, however, that the U.S. system has worked better recently than the Soviet system, and this evidence is playing a heavy role in discussions of science in post-Soviet Russia.

Russian Debates over the Organization of Science

When, under Gorbachev, critical voices were at last permitted to be expressed in the media, a chorus of calls for changes in Soviet science emerged.

In the last years of the USSR's existence a strong reform movement developed among some scientists, especially among young researchers in the fundamental sciences. These junior scientists were well aware of the deficiencies of the hierarchical Soviet scientific system, and they could not understand why a few hundred members of an honorary learned society, the Academy of Sciences, should be given administrative authority over almost all fundamental science. Most of these critics of the existing scientific establishment also favored democratic reform of the Soviet government. In fact, they were far more unanimous in their opinions about how the *government* should be organized than they were in their views about how *science* should be organized. As a result, they were never able to form an effective political coalition on matters of science policy. They did not have clear answers to crucial questions such as: "As the Soviet Union makes the transition from totalitarianism to democracy, how much of the science establishment needs to be changed in step with the political changes sweeping the country?" "To what degree did science in the former Soviet Union display the characteristics of the centrally controlled political and economic system that had formed and nurtured it?" "What is the way that science should be organized so that scientists are most productive and creative?"

The aspiring reformers split into different groups in their answers to these questions. Some of them favored what was often called "the American system," which meant that they believed that the seat of fundamental science should be in the universities, not in institutes of the Academy. Some of the more radical of these critics proposed giving the research facilities of the Academy to the universities, although there was little agreement or clarity on just how this could or should be done. Others believed that preservation of the institute system was necessary, but favored separating these institutes from the Academy of Sciences, which would become a mere honorary society without administrative authority. In this scheme, the institutes that had previously been in the Academy would be reorganized into an association of research laboratories, perhaps like the CNRS in France or the Max Planck Gesellschaft in Germany. Still others favored keeping the basic organizational structure of the old Soviet system of science but proposed to "democratize" it by including in the major decision-making bodies of the Academy—the General Assembly and the Presidium—elected representatives of the scientists throughout the Academy, including the junior scientists. Opposing all three of these blueprints for the future were the top officials of the old Academy of Sciences of the USSR, who struggled mightily almost up to the collapse of the Soviet Union itself to keep the old all-union structures, along, of course, with their offices and privileges. They pointed out that whatever the flaws of the Academy of Sciences of the USSR, it nonetheless possessed the best scientists in the country. They feared that a radical reform of Russian science would destroy the strongest institutions.

At the peak of the wars over science before the collapse of the Soviet Union, which came in the period 1989–1991, many different positions emerged. Everybody, it seemed, had grievances about science. Researchers in fundamental science chafed at governmental and political controls and at the preferences often given to topics of industrial and military importance. They wanted less interference and more money for fundamental science.[29] Soviet government bureaucrats and economic planners from GOSPLAN (State Planning Commission) considered these pure scientists unaccountable and arrogant; these government administrators pointed out that the USSR spent a higher percentage of its gross national product on science than any other country, in fact, more than all West European countries taken together, and one and a half times as much as Japan. Yet Soviet scientists were far less productive than Western and Japanese researchers; according to one GOSPLAN official, the average Soviet scientist was four times less productive than the average U.S. scientist.[30]

Scientific researchers in the Academy of Sciences, especially the junior ones, complained that they were entirely excluded from the influential policy-making bodies of the Academy. They called for a more democratic system of administration within the Academy.[31] Corresponding members of the Academy of Sciences resented their junior status and wished to be full members; they pointed out that in other countries, a two-rank system of membership in academies and scholarly societies was rare.[32] Engineers resented their expulsion from the Academy in the 1960s and called for a new Academy that would include them.[33] University professors were unhappy about the lower status of university science relative to that in the Academy and called for a transfer of science to the universities; if such a transfer was not possible, they wanted the rank of "academician" for themselves.[34] Russian nationalists were irritated by the absence, in the Soviet system, of a Russian Academy of Sciences while all the other republics in the Soviet Union had their own academies.[35]

Political reporters and outside observers were critical of what they considered the arrogance and philistinism of the leading administrators of the Academy of Sciences of the USSR. A particularly offensive symbol of their expensive bad taste was, in the critics' opinion, the garish new skyscraper intended for the presidium of the Academy being constructed near Gagarin Square in Moscow; this building would require a service staff of 600 people, and its offices were originally slated to be entirely those of administrators. In the near-revolutionary situation that was developing in science, the building served the role of Marie Antoinette's diamond necklace before the French Revolution, inflaming opinion against the old regime.[36]

The deputies in the increasingly democratic parliaments of Russia resented the financial unaccountability of the Academy of Sciences to the government; the Academy always wanted more money, and it wanted absolutely no interference in how it spent the money. The ire of the deputies was inflamed by Mikhail Gorbachev's August 1990 decree making the Academy fully independent of government influence, and able to possess its own property, while remaining totally dependent on the government for its budget. (This decree was later reversed by Boris Yeltsin.)[37]

Provincial scientists, those outside Moscow, St. Petersburg, and Novosibirsk, resented the "big science" centered in those cities. They wanted the status awarded scientists in the Academy, most of whom were in Moscow. These provincial scientists were the driving force behind the creation of a rival Russian Academy of Sciences in 1990 and 1991, and they also created a dozen or more rival "academies."[38]

In the last years of the Soviet Union, from 1987 to 1991, dozens, if not hundreds, of angry articles were published in the Soviet press about science. In addition many groups of scientists organized meetings at which resolutions were drafted calling for dramatic changes in the organization, politics, and financing of Russian science. Since this book concentrates on the post-Soviet period, we cannot explore this fascinating period of time adequately here. The topic deserves a separate doctoral dissertation or monograph, and the archival information necessary for writing such a work exists both in the United States and in Russia.[39]

In order to give a better taste of the spirit of the debates about science in the last months of the old Soviet Union, we will discuss here only several articles out of the multitude of possibilities.

One of the most devastating critiques of the Soviet Academy of Sciences was written by a man who knew what he was talking about, Academician Vitaly Ginzburg, a very distinguished physicist (and later a Nobel laureate).[40] Ginzburg revealed how the elections to membership in the Academy—supposedly an independent body electing its members without political interference—were manipulated by the Soviet authorities. Ginzburg told of "the telephone law" whereby Communist Party leaders would call the president and top officers of the Academy and tell them who they wanted to win election to membership and who they wanted to fail. In order to make sure that their preferences were followed, the elections were held in such a way that it was clear who obeyed Party dictates and who did not. When voting for new members, the existing members were given ballots on which for each candidate there were columns after the candidate's name marked "elect" or "vote down." However, the established system was that if a member wanted to vote "elect" for a given candidate, he merely folded the ballot and dropped it in the ballot box, making no mark upon it. In order to "vote down" a candidate the member was forced to pull out his pen and mark the "vote down" column. This was done in the open auditorium with many other people, including Party observers, present. Watching who, if anybody, pulled out their pens and made marks on the ballots was something the Party observers kept track of. It was absolutely clear who voted against the Party-controlled slate, and the clear implication was that dissidents might face retribution. By taking advantage of Gorbachev's policy of *glasnost'* (openness) and telling this story in a leading newspaper, Academician Ginzburg revealed that elections to the Academy were often a farce.

In another article, titled "Is this an Academy or a fortress?" Academician V. Tishkov lambasted the Academy as a "grandiose ministry of science" dedicated to preserving its own power and privileges and impervious to all real calls for reform.[41] Tishkov, the director of the Institute of Ethnography, observed that during the last fifty years in the "developed countries of the world" science had created powerful research universities where the Soviet-style division of fundamental science, applied science, and teaching had no place. He castigated the administration of the Academy as immoral and obscurantist and called it the home of "hypocrisy, fawning, lack of principle, and protectionism."

In another article Academician B. Raushenbakh severely criticized what he called "monopolism" in Soviet science, the situation in which one scientist could force his views upon every one else. He said that the "institute system" of the Academy facilitated this authoritarianism. Many members of the Academy, Raushenbakh continued, practiced "gigantomania," the creation of enormous institutes which they directed and in which hundreds or thousands of researchers were forced to follow their dictates. No wonder, he observed, that in the West university science departments, which might harbor only a dozen or so members but these were free to get outside grants to pursue their own ideas, often outperformed institutes of the Soviet Academy with hundreds of researchers following the dictates of a single powerful director. Yet another researcher in the Academy put forth the same message in an article titled "O Give Us Freedom!"[42]

In still another article, "Why does our science nurture academicians while foreign science nurtures Nobel Prize winners?" a scientist from Novosibirsk, A. Burshtein, answered his own question by saying, "The administration of our science is extremely centralized, incompetent, and selfish."[43] He called the administration of the Academy "feudalistic." Burshtein's recommended answer to the problem was to give funds to researchers in the Academy "bypassing" all levels of the Academy administration: the president, the presidium, the departments, and even the directors of the institutes. Burshtein wanted a system of grants based on peer review by "elected scientists." He did not have a very clear idea of just how this should be done, but he thought scientists should have "direct access" to government funds. Burshtein did not dream that within two years after he wrote his militant article the Soviet Union would have disappeared and almost no funds would be available for science at all.

The position of the old guard, the supporters of the status quo, was enormously weakened by the events surrounding the abortive putsch in August 1991. When army generals and police officials attempted to reverse Gorbachev's reforms and restore authoritarian, if not totalitarian, rule, the top leadership of the Academy of Sciences silently sided with them. The Academy's leaders were, in fact, along with the conservative heads of the Writer's Union, among the very few intellectuals in the dying Soviet Union who sided with the old order. The leaders of the Academy of Sciences of the Soviet Union believed that their fate was inextricably linked to the USSR itself. When the coup failed and its leaders were arrested, the leadership of the Academy was thoroughly discredited. *This* was the moment when basic reform seemed possible.

In the immediate aftermath of the coup, the old Academy was in a very precarious position. The Soviet Union was breaking up into its constituent republics, and already a rival "Russian Academy of Sciences" had arisen. In its initial conception, this new academy allowed no place for the old system; according to the "basic principles" of the new academy published in the press: "The Academy of Sciences . . . does not contain scientific-research subdivisions or schools" and "Material compensation for membership in the Russian Academy of Sciences is not provided."[44] If this model had been followed, a radical reform of Russian science would have occurred, moving Russia toward either a university-based research system or an institute-based one that was separate from the Academy of Sciences. According to these first principles, the new Russian Academy of Sciences would become a mere learned society, prestigious but not administratively important. However, this transition did not occur. As Aleksei Zakharov, a leader for the reform movement, observed, "An historic opportunity for a constructive reformation of the entire organization of Russian fundamental science was missed."[45] What happened?

What happened was that the allurements of the old system were too enticing even for many of the reformers to resist. Many scientists had been corrupted by the old system and preferred its continuation. As Zakharov continued, the people organizing the new Russian Academy of Sciences "faced a choice: lead an academy without privileges and institutes, or be rank-and-file members of an academy with institutes, salary supplements and privileges. They chose the second option. Which would you choose?"[46] In this stark question we see the error in the position of those who maintain that the Soviet system of scientific research was imposed on unwilling scientists. If giving up the Soviet system meant giving up their salaries and privileged positions, many Russian scientists understandably opted for the old system.

A way to make the transition to a new independent Russia without giving up the state-socialist system of fundamental research was found by fusing the two academies, the old Academy of Sciences of the USSR and the new Russian Academy of Sciences; the approximately 300 full members and 450 corresponding members of the "big" Academy were combined in late 1991 with the newly elected 39 full members and 108 corresponding members of the just-born one to form a single Russian Academy of Sciences (RAN), which took over the research establishment of the old Academy. A part of this agreement was the abandonment, at least for a time, of the idea of true reform that had motivated many of the early advocates of the Russian Academy of Sciences. Instead of becoming a learned society in the West European or North American mode, the Russian Academy of Sciences remained a vast research bureaucracy of laboratories and institutes claiming primacy in fundamental research.

The preceding pages describe the period in the last years and months of the old Soviet Union, which could be described as the "radical and somewhat naïve" period of the discussions of reform in late Soviet science. All the participants in this debate assumed that science in the new Russia would remain well funded, just as it had been in the Soviet Union. They did not know that a great financial crisis was coming after the collapse of the Soviet government, in which the fundamental question facing Russian scientists was not how they should reorganize their science establishment, but whether Russian science would survive at all.

(It is easy now to say that Russian natural scientists "should have" known that an economic train wreck was coming, but the fact is that most of them did not. They were concentrated on their laboratory work and they had little knowledge of economic trends.) In the desperate years coming, the positions taken by the many disputing groups of scientists in the last months of Soviet rule appeared, in retrospect, indulgent and unrealistic. But the main issue of those heady days—how should Russian science be organized and financed?—was only postponed, not eliminated. After the Russian economy began to recover in the late 1990s and the early years of the next millennium, this question would return, rephrased as "How should Russian science be organized and financed in a globalizing world in which Russia's main competitors are democratic, free-market, innovative societies?" No longer, however, did anyone hope for a "revolution" in the organization of Russian science; even those scientists most fervently wanting organizational change could only hope for gradual reform, and even that would be elusive.

2 Breakup of the Soviet Union and Crisis in Russian Science

Science in Crisis

What were the causes of the deep crisis in which Russian science found itself after the dissolution of the Soviet Union in December 1991? Following the disintegration of the USSR, a rapid decrease in the status of science and the prestige of research work occurred, both among policy-makers and among the general public. The effects of this decrease on the scientific establishment were profound and widespread

The Effects of Financial Decline

Science had been strongly supported by Soviet ideology, and with the discrediting of that ideology, it moved much lower in social and political priorities. With the disappearance of the central all-union federal budget, expenditures on science were greatly reduced. In the early post-Soviet period, changing the political, social, and economic systems was thought to be much more important than changing or helping science. Under financial constraints, science was seen as a luxury. And if one takes into account that at the end of the 1980s about 97 percent of the support of science came from the federal budget, it becomes obvious how dramatic the situation was.[1]

Moreover, one of the directions of the early reforms was demilitarization; as a result, defense research and development quickly lost support. The effect of such a rapid change was traumatic because about three-quarters of the total research and development complex was related to defense.

At the same time, attempts by researchers to protest against their new plight were not taken seriously because of what appeared to be greater emergencies elsewhere. The strikes of miners and the dissidence of the military forces were much more dangerous for the power structures than the protest meetings of scientific researchers. The science and technology sphere was pushed into the background, and the government did not even try to maintain minimum living standards for researchers.

As a result of the liberalization of the economy upon the disappearance of Soviet controls, prices grew in Russia in 1991–1992 at a hyperinflationary rate. In 1992 prices rose, according to the Russian government, at a rate of 1,500 percent; according to independent experts the rate was 2,200 percent.[2] Average real incomes contracted to 50 percent of the level of 1991 (see table 2.1).

Table 2.1. Socioeconomic indicators of Russia (in percent of previous year)

	1991	1992	1993
Gross domestic product	95.0	85.5	91.3
Capital investments	85.0	60.0	88.0
Disposable income	121.1	50.0	116.0

Source: Russian State Committee on Statistics.

Because of the fact that many other economic problems seemed more pressing, science was excluded from top-priority needs and began to be financed on the principle of "whatever was left." The problems at the center of attention of the government—achieving economic stability, halting the decline of manufacturing, launching irreversible reforms—were ones that science did not seem to be an instrument for solving. As a result there was a collapse of government support for science (see table 2.2). The problem was deepened by the fact that the administration of science remained the administrative-command scheme of the Soviet period. As before, the organization and financing of research ran along governmental lines; there was total governmental control, and restricted contact with the world scientific community.[3]

The level of state support sharply diminished both for civilian and defense science: in 1992 the total expenditures on science from all sources decreased by two times that of the previous year. Suddenly, in 1992, Russia fell significantly behind other developed countries in terms of expenditures on science (table 2.3). By 1994 the level of financing of Russian science was almost six times lower than in developed countries of the West.

If one takes into consideration the fact that the size of the gross domestic product (GDP) also constantly contracted during this period, the real level of financing of science in Russia in 1994 was 20 percent of that of 1991. Since in this crisis situation the first priority of administrators of scientific institutes was the support of their personnel, there was practically nothing left in their budgets for purchasing instruments and equipment, or for libraries or electronic information services. This led to the rapid obsolescence of the scientific establishment. By January 1, 1993, 38.7 percent of the instruments and equipment used by researchers was more than eleven years old. In computer technology half of all the equipment had been in service more than five years.[4] (Obsolete equipment was a problem in the Soviet period as well, but certain privileged institutes, such as the Lebedev Institute of Physics, the Kurchatov Institute of Atomic Energy, and the Joint Institute for Nuclear Research in Dubna, were well supported with equipment. After the fall of the Soviet Union, even these institutes lost equipment support. Unfortunately, we do not have reliable comparative date on age of scientific equipment in both the Soviet and post-Soviet periods.)

The redirection of funds toward salaries and basic utilities did not save the

Table 2.2. Basic indicators for science in Russia

	1991	1995	1998	1999	2001	2003
Appropriations for R&D from the federal budget in percentages of GDP	1.85	0.54	0.40	0.50	0.54	0.71
Total expenditures on R&D from all other sources in percentages of GDP	2.43	0.85	0.93	1.06	1.24	1.25
Share of the entrepreneurial sector in total expenditures on R&D, %	0.0	24.1	22.8	22.6	24.8	22.8
Number of researchers per 10,000 of population	108	60	77	78	78	75
Number of researchers as percentage of previous chronological year (not previous column)	88.5	98.7	91.9	100.8	99.1	98.8

Sources: *Nauka Rossii: 1994. Statisticheskii sbornik,* TsISN, Moscow, 1994, pp. 95, 105; *Nauka Rossii v tsifrakh: 1996. Statisticheskii sbornik,* TsISN, Moscow, 1996, pp. 34, 85; *Nauka Rossii v tsifrakh: 2004. Statisticheskii sbornik,* TsISN, Moscow, 2004, pp. 46, 70, 71, 75.

situation. There simply were not enough funds even for such essentials. The living standard of scientific researchers fell by five or six times, and by 1993 56 percent of scientific researchers received salaries lower than the subsistence wage. An abandonment of scientific careers began among researchers, a brain drain both internal and external.

Meanwhile, the scientific literature and knowledge available to researchers in Russia diminished in quantity. Subscriptions to journals, especially foreign journals, lapsed, and attendance at international conferences dwindled. Deprived of subsidies, scientific presses dramatically cut their publication activities. The output of the publishing house Nauka, the press of the Russian Academy of Sciences, dropped by 1995 to about one-fifth of its previous level. By the mid-1990s it was not unusual to find scholarly journals in the former Soviet Union that instead of publishing the usual six or twelve issues a year published only one or two, perhaps labeling each copy "Issues 1–6," indicating that six issues were combined in one.

In 1989 out of each 10,000 actively employed people in Russia 130 were scientific researchers; by 1995 that indicator had fallen to 60. The rapidity of the decrease in science was two times greater than in other branches of the economy. In 1991 the contraction of scientific researchers was 13.7 percent; in 1992, 21.1 percent; in 1993, 32.2 percent; in 1994, 43.1 percent. The highest

Table 2.3. Governmental science expenditures in Russia and several Western countries

Country	Share of expenditures on science as % of gross domestic product		
	1992	1993	1994
Russia	0.57	0.52	0.44
United States	2.7	2.6	2.5
Japan	2.8	2.7	2.8
Germany	2.5	2.5	2.5
France	2.4	2.4	2.4
United Kingdom	2.2	2.2	2.3

Sources: 1. Russian State Committee on Statistics; 2. Science & Engineering Indicators—1996, National Science Foundation, Washington, D.C., 1996, appendix table 4-33, p. 154.

rates of contraction were in university and industrial sectors of science, not in the Academy of Sciences sector (see table 2.4).

The official statistics on research personnel disguise a large number of people who by 1993 were no longer really working at research institutes but still received their salaries. By doing so they kept their labor books at the institutes and ensured their pensions. Titular employment in the Academy gave them prestige wherever they worked. Institute directors often went along with this practice because their very meager budget funds depended in part on the number of employees.

Simultaneously, beginning in 1990 there was a decrease in the number of graduate students. The worst year in this respect was 1993: in that year there were fewer graduate students than in any of the immediately preceding or subsequent years. In 1970, for every 1,000 students in universities there were 107 graduate students; by the beginning of the 1990s the number was 66; in 1992–1993 several universities and institutes of the Russian Academy were forced to liquidate their graduate programs.

Impact of the Crisis on One Institute

One can get a better idea of the dimensions of the science crisis and its effects on institutions by looking at one well-known institute, the Ioffe Physico-Technical Institute in St. Petersburg. This institute is a part of the Russian Academy of Sciences and for many years has been directed by Zhores Alferov. It is one of the most famous physics institutes in all of Russia and has often been called "the cradle of Soviet physics." Four of its researchers have won Nobel Prizes: P. L. Kapitsa, L. D. Landau, N. N. Semenov, and Alferov himself (in 2000). The leader of the Soviet atomic project, I. V. Kurchatov, was also a young scientist at the Ioffe Institute, as was Iulii Khariton, another prominent scientist in the atomic project. Although the record of the Ioffe Institute in re-

Table 2.4. Rates of change in the number
of researchers in different scientific institutions

Scientific institution	Rates of change in personnel, % of 1991 situation	
	In 1992	In 1993
Russian Academy of Sciences	+0.4	–6.5
Industrial S&T	–9.4	–31.0
Higher Educational Institutions	–15.2	–35.4

Based on *Nauka v Rossiiskoi Federatsii*, Goskomstat RF. Moscow, 1995, p. 13.

cent years has not been quite as stellar as it used to be, it is still an outstanding center of physics research. Yet, as we will see, its great reputation did not protect it from deep crisis in the early and mid-nineties.

In 1990, the Ioffe Institute had a staff of 3,000, of whom 1,400 were scientists. By January 1995, the staff had dropped to 2,500, of whom 1,200 were scientists. In March 1995, approximately 60 of these 1,200 were working abroad. Summing up the situation with personnel, the number of research scientists in the institute had dropped only by about 15 percent, or possibly 19 percent, if the scientists who went abroad did not return. Although this is a significant decrease, it is not nearly commensurate with the decrease in the overall budget of the institute, which had dropped much more dramatically.

Between 1990 and 1995, the budget of the institute decreased by about fifteen times. In rubles adjusted for inflation, in 1990, the budget of the institute was about 66 million rubles; in 1995, it was 3.4 million rubles. How could the institute survive in the face of such draconian financial cuts? Its director, Alferov, decided to dramatically redistribute the remaining budget, concentrating on maintaining salaries at the expense of almost everything else.[5] In 1990, only about 7 percent of the institute's budget supported salaries; 45 percent of the budget went for materials and equipment. In 1995, almost 70 percent of the institute's budget was allocated to salaries; only about 5 percent went for materials and equipment. We can sum up the situation in the following way: the Ioffe Physico-Technical Institute was surviving by ceasing its purchases of materials and equipment and devoting its meager resources primarily to the payment of its staff. Such a stopgap policy could obviously work only for a short while. For a few years it was possible to stop buying instruments and materials, but if this pattern continued for any length of time, the laboratory would fall hopelessly behind. New equipment—computers and instruments of all kinds—are absolutely essential in modern scientific research.

By the middle and late 1990s, the pattern in the Ioffe Institute was being repeated all over the former Soviet Union. Research institutes were trying to hang on to their research staffs, and in order to do so they were stopping almost all other expenditures. The overall staff of the Russian Academy of Sciences decreased only 19 percent during 1990–1995, while the budget decreased by almost 1,000 percent. (Even today, after the passing of the great financial crisis of the mid-nineties, most of the government funds going to research institutes are designated for salaries; as a result, the administrators of the institutes often look elsewhere for money for equipment, such as foreign grants.)

The "Brain Drain"

The brain drain constitutes a revealing indicator of the crisis in Russian science. The drain of talent was of two kinds: foreign emigration and internal outflow to other branches of the economy. According to specialists in the study of brain drain, the ratio between internal outflow and external emigration in 1992–1993 was ten to one. In that year physicists and mathematicians made up about half of the total emigration, internal and external. Especially alarming was the fact that a large fraction of those leaving science comprised either the scientific elite or young researchers. Middle-aged scientists were more inclined to stay in their old positions, even if in desperate conditions.

Some of the reasons scientists wished to leave, especially in the early 1990s, were political or ethnic. Large numbers of Jews left for such reasons, and many Germans also left. "Science mobility" may be an appropriate description of the reasons many other scientists left: top scientists from most countries are highly mobile, and many of them seek to work where the leading specialists in their fields are located. But probably the most important reason for departure was material—the desire of scientists to improve their financial situations.

The outflow of scientific researchers strongly changed the age structure within science: in 1993, among researchers in fundamental research only 4 percent were young (under thirty years of age); in applied research, 12 percent; in pilot-plant and initial production phases, 9 percent. As a result of the departure of younger scientists through brain drain, the average age of researchers in Russia rose. In 1994 the share of researchers who were under twenty-nine years of age was 9.2 percent, and the share of those who were aged thirty to thirty-nine was 24 percent; by 1998 those numbers decreased to 7.7 and 18.1 percent, respectively.[6]

The problem of brain drain is very complicated to capture statistically. There have been no all-Russia comprehensive surveys of this phenomenon, only a number of partial surveys with quite different results. The most conservative estimates give the outflow abroad (both for permanent residence and for long-term contracts) at about 7,000 researchers for the years 1993–1996, and the least conservative estimates give the outflow for the same period at between 30,000 and 40,000 researchers. The Russian Ministry of Science produced data on external brain drain (for permanent emigration, not temporary residence abroad for contract purposes), stating that between 1991 and 1996 there was a rather stable outflow of about 2,000 researchers a year. In the period 1997 to 2000 this average indicator decreased substantially to 1,200 to 1,400 researchers each year.[7] In 2001 and 2002 the number decreased again to approximately 500 to 600 researchers a year. After 2003 this statistic was not collected.[8]

Internal brain drain, that is, leaving science for other types of activity, is even more difficult to estimate because it is impossible to collect accurate data. There was a large share of hidden internal emigration caused by researchers continuing to list themselves as staff members at a given institute but actually spending most of their time in activities outside science. According to anecdotal evidence, only about 10 percent of all researchers in the 1990s were working full time in the science and technology sphere. Many of them were forced to seek part-time employment elsewhere in order to support themselves and their families.

But even the relatively scarce data we possess reveal that the process of external brain drain has changed during the last fifteen years. By analyzing data from different sources, we conclude that four waves of brain drain can be seen in post-Soviet Russia. Of course, these waves were not absolutely distinct, and our identification of them is only an approximation.

The *first wave,* from the end of the 1980s to the early 1990s, consisted largely of elite scientists who were well known in the international scientific community. About 70 percent of the researchers who left Russia in these early years continued their scientific careers at universities and R&D organizations abroad. Another significant outflow at this time was ethnic emigration, especially by Jews and Germans. Germany and Israel accepted about 80 percent of all emigrants in these years, but in many cases these newcomers later moved to the United States, seeking more favorable research conditions and immigration laws.[9]

The *second wave* of emigration came in 1992–1993, the time of the most intensive outflow. But only 20–40 percent of the total number who emigrated abroad at this time stayed in science after they arrived in foreign countries.[10] During these years, many scientists in Russia also emigrated internally, permanently leaving science for other activities.[11]

The typical emigrant abroad between 1992 and 1993 was a thirty-one- to forty-five-year-old man with a doctorate, engaged in theoretical research, often having published widely.[12] According to different surveys, in terms of scientific disciplines, physicists and mathematicians composed more than 50 percent of the total number of emigrants, followed by biologists (about 30 percent) and chemists. The biggest share of emigrants departed from Moscow, St. Petersburg, and Novosibirsk—the major Russian scientific centers. The distribution of emigrants by countries showed that the first destinations (in descending order) were Germany, Israel, and the United States.

The *third wave* took place from 1994 to 1998. The typical representatives of this wave were biologists (especially in fields such as genetics, molecular biology, and virology), computer science specialists, and programmers. There were fewer physicists, mathematicians, and chemists in this wave, mostly because of diminishing demand for these specialists in foreign countries, especially in the United States. In 1997–1998 external outflow was stable at about

1,000 researchers per year leaving permanently, with an additional 4,000–5,000 researchers per year working abroad on a contract basis. Graduate students and young researchers were prominent among those leaving during these years, resulting in a scarcity of young (thirty to forty years of age) researchers in Russia.[13] During 1996–1998 the number of graduate students from the former Soviet Union (FSU) in U.S. universities increased by 29.5 percent. Russian graduate students abroad outnumbered undergraduates.[14]

In 1999 a *fourth wave* of emigration started, with the number of emigrants increasing to 1,400 in this year. According to a 2000 survey conducted among researchers from different fields, the major reason given for emigrating at that time were (in descending order by the number of researchers citing them): (1) low levels of salaries for both research and teaching; (2) ever-worsening shortages of equipment and instruments for conducting fundamental research; (3) absence of prospects for career growth in Russia; (4) low prestige of scientific careers in Russia; and (5) an unstable political situation. Low salaries were cited by more than half the respondents; each of the other reasons was cited by approximately 30 percent.[15]

Characteristic features of this fourth wave were departure from Russia after first defending a thesis there, and "pendulum migration," a growing number of researchers who move back and forth rather than leave Russia permanently.[16] Western scientists, especially Americans, often strengthened these tendencies by searching for future post-doctoral candidates among graduate students they met while visiting Russian laboratories. In turn, Russian researchers increasingly sought places in foreign laboratories while they were still in graduate study in Russia. Young researchers had the strongest motivation for emigration, often encouraged and inspired by their older colleagues. As surveys in this period showed, 50 percent of researchers in the older generation thought that young scientists should go abroad to work temporarily, and 9 percent believed that such young people should leave Russia forever; only 21 percent thought that these youths should try to pursue scientific careers in Russia.[17]

Demography and Personnel Problems

Personnel problems have existed in Russian science throughout the post-Soviet period. The most troubling have been changes in the demography of the scientific community resulting from the constant departure of science students and young scientists, and the "washing out" of middle-aged scientists. The aging of the scientific community therefore continued. As table 2.5 indicates, there has not been a close correlation between the number of scientific workers and the financial indicators. Increases in investment in science do not, at least so far, seem to result in increases in the number of researchers. With the exception of a short period from 1999 to 2000, when a slight growth in the number of researchers occurred, this number has constantly declined. This exception was caused by the economic crisis in August 1998, when a number of people who had earlier been researchers returned to their old jobs to wait out

Table 2.5. Age distribution of Russian researchers, %

Year	29 years and younger	30–39 years	40–49 years	50–59 years	60 years and older	Total
1994	9.2	24.0	31.7	26.1	9.0	100
1998	7.7	18.1	28.3	27.9	18.0	100
2000	10.6	15.6	26.1	26.9	20.7	100
2002	13.5	13.8	23.9	27.0	21.8	100

Source: *Nauchnyi potentsial i tekhnicheskii uroven' proizvodstva*, Ministerstva obrazovaniia i nauki RF, RIEPP, Izdatel'stvo RUDN, Moscow, 2004, p. 22.

the difficult times. But soon the previous downward trend resumed, although at the present time it has slowed somewhat.

Nonetheless, there has been an increase in the percentage of young scholars (among all scholars) in recent years. How significant this improvement is remains unknown since quite a few of the young researchers, according to surveys, hope to leave; at the same time, the middle group (age 30–39) continues to decline.

Table 2.5 shows that there has also been a decline in the cohort of scientists from 40 to 49 years in age. The overall number of researchers diminished from 1998 to 2002 only by 0.5 percent, but the absolute number of researchers in the age group 30–39 diminished by 25 percent, and those from 40–49 by 16 percent. At the same time the growth of the number of scientists older than sixty years was 20.4 percent. The most alarming situation among researchers occurred in the institutes of the Russian Academy of Sciences, where, according to official statistics, the average age of researchers was almost five years greater than the average age of scientists throughout the whole country.

To understand better the unusual demographic situation among scientists that developed in Russia, it is useful to compare the age structure of the scientific community in Russia with that in the United States (based on statistics for the years 1999–2000).[18] Among established working scientists the biggest differences between the two countries were in the young and the old. In between, at about age fifty, the two countries had about the same distribution. However, the age group 30–39 comprised about 30 percent of all U.S. researchers, while only about 16 percent of all Russian researchers were in this group. On the other hand, only about 6 percent of all U.S. researchers were older than 60, while this group of Russian researchers comprised about 21 percent of the total. In other words, in Russia the scientific community was significantly older than in the United States.

These statistics conceal the fact, however, that among the very young (younger than twenty-nine, including many graduate students), the distributions in the two countries are rather similar (15 percent of U.S. scientists fall into this group and 11 percent of Russians). What this latter statistic reveals is that many young

Table 2.6. Number of graduate students and those finishing graduate studies in Russia

	1995	1998	1999	2000	2001	2002	2003
Number of graduate students	62,317	98,355	107,031	117,714	128,420	136,242	140,741
Number who finished graduate studies	11,369	17,972	21,982	24,828	25,696	28,101	30,799
Share of those who finished graduate studies out of all those in graduate studies, %	18.2	18.2	20.5	21.1	20.0	20.6	21.9

Sources: *Nauka Rossii v tsifrakh—2000: Statisticheskii sbornik,* TsISN, Moscow, 2000, p. 23; *Nauka Rossii v tsifrakh—2004: Statisticheskii sbornik,* TsISN, Moscow, 2004, p. 36.

Russians enter science only temporarily. They engage in research up to the moment they are able to find higher-paying jobs elsewhere in the country, or until they emigrate abroad, where they may or may not remain in science.

At the same time, statistics on the number of graduate students and those who finish graduate study show a constant growth in the number of young people attempting to receive the degree of "kandidat" of science (often compared to the Ph.D. in the United States). These statistics appear promising, since they demonstrate the internal potential in Russia for the inflow of young researchers into science. However, upon closer examination of the statistics the situation looks less optimistic. Table 2.6 gives the relevant information.

From table 2.6 we see that the number of graduate students who finish study is small compared to the total number studying (only about 20 percent). Before the disintegration of the USSR this ratio was somewhat higher; from 1985–1990 it was 26 percent.[19] For comparative purposes, studies in the United States indicate that the attrition rate for Ph.D. programs is 40 to 50 percent.[20]

Unfortunately, the increase in the number of graduate students and young scientists does not, in itself, prove a growth of interest among young people in scientific careers. Graduate study in post-Soviet Russia serves several other functions; it is a way to avoid serving in the army, a way to delay professional choices, and a way to prepare for a career in business or other pursuits. Social surveys of graduate students in Russia indicate that no more than 20 percent plan to continue scientific work (anywhere, either in Russia or abroad) after they have completed their studies.

Anticipated material rewards are often decisive; surveys indicate that the single factor most likely to persuade young people to stay in science is a high base salary. The size of this desired base level varies in different regions of the country, but on average it is about $500 a month. However, other factors also play a role.

Sociological surveys made by one of the authors in 2001–2002 in 21 regions of Russia revealed the basic reasons that young people often do not want to become professional scientists.[21] Included in the surveys were students at all levels working and studying in a great variety of specializations. The surveys covered 1,400 undergraduates, 450 graduate students and 1,200 young scientists.

The surveys showed that already in the university, during their undergraduate years, two factors were at work that caused the students' disappointment in research activity: the obsolescence of equipment in the universities and the absence of interesting scientific tasks. At the same time, the students' respect for science as an activity still continued and their desire to emigrate was not high. Later, the significance of poor equipment and inadequate intellectual stimulation continued to grow (among graduate students and young scientists), leading finally to the departure of some of the young people abroad and/or into other forms of employment. Among graduate students the interest in emigration was higher than among undergraduates, and the possibility of work abroad became one of the main motivations for doing graduate work.

The surveys also showed that the possibility of keeping young people in science was high if certain changes were made, such as improving equipment, invigorating the research process, and paying better salaries. Another obstacle to young people is the inflexibility of the existing organizational structure of science, which produces a very slow turnover of personnel. It is difficult for graduate students and young scientists to find a permanent place in research institutes and universities, especially in capital cities. Furthermore, a system of post-doctoral study (post-docs), such as exists in many other countries, is only beginning to be established in Russia. For this reasons and for the other reasons already named, one of four surveyed scientists thirty years old or younger expressed the wish to leave for work abroad. These young people are understandably interested in their professional prospects, and they see small chance in Russia of becoming leader of a laboratory or a distinguished scientist (e.g., "leading researcher," etc.)

Attempts to Cope with the Crisis

As described in the previous chapter, the radical calls for changes in the organization of science in the last months of the old Soviet Union failed. After the collapse of the USSR, the "revolutionaries" passed off the scene and the "evolutionaries" gained the upper hand. An important figure in this evolutionary approach was Boris Saltykov, who served Russia as minister of science from November 1991 to August 1996. Saltykov and the members of his team were very aware of the defects of the old Soviet system of organizing science, and they favored strengthening the universities relative to the Russian Academy of Sciences, but they also understood that these changes must happen gradually. Instead of calling for radical reforms, they attempted to cope with the crisis on the basis of the old system of organization while hoping for bigger changes in the future. Leaders of the Academy joined in support of this approach; conservatives and moderates were able to find agreement on the principle that old systems should not be thrown out before new ones were developed.

The Academy of Sciences, which had been in such a precarious position in the last months of the Soviet Union—discredited in the minds of many scientists and political observers, and bereft of financial support as the federal budget disappeared—found a new lease on life, even though it continued with very little money. After the old Academy of Sciences of the USSR was combined with the new Russian Academy of Sciences in late 1991 and early 1992 the new "Russian Academy of Sciences" that emerged possessed all the research institutes of the old Soviet one (except, of course, for the institutes in the newly independent states), and Russia continued to have a research establishment that was different in organizational principles from the scientific establishments of most countries in Western Europe and North America and rather similar to that of the old Soviet Union. The new head of the Russian Academy of Sciences, Iurii Osipov (he is still the head today), wanted to keep a dominant Academy but also favored moderate reforms, such as the election by junior researchers of representatives in the General Assembly governing the Academy. This latter political reform was accomplished, but it had almost no effect. The representatives of the junior researchers did not gain significant political influence. They were not permitted to vote on the most important matters, such as elections of members. Many of them hoped some day to be elected either corresponding or full members of the Academy themselves, and therefore they did not usually irritate their seniors by causing trouble.

Science administrators in Russia found unofficial ways of coping with the difficult financial situation in Russian science in the nineties. Many institute directors—especially those who headed institutes in desirable locations, such as in or near Moscow and St. Petersburg—found that renting out institute space, often to commercial enterprises, was one way to increase revenues. Another approach was to encourage researchers to remain on the staffs of the institutes while receiving their salaries from outside activities. Such former researchers were sometimes referred to as "ghost members." Although these unofficial means of coping with financial constraints significantly relieved budgetary pressure on institute directors, they also clearly detracted from the active research output of the institutes.

In the nineties the most important events in Russian science were the infusions of aid that came from foreign foundations and organizations, and the creation of new science foundations in Russia itself. These developments will be discussed in detail in the following chapters.

The Beginning of Improvement

One of the notable events in Russian science in 2000 was the award of a Nobel Prize to physicist Zhores Alferov, the director of the Ioffe Physico-Technical Institute and a member of the Russian Academy of Sciences. The last time a Nobel Prize was previously awarded in the natural sciences to a citizen of the Soviet Union or of Russia was in 1976. Although Alferov's award was made for work done in the Soviet period about twenty years earlier, it was an interesting coincidence that the prize was given to a Russian researcher in the same year that signs of recovery in Russian science began to appear. The award of the prize to Alferov was psychologically important to Russian science, signaling that the worst period of the crisis was over.

At the end of 1999, two other significant events occurred. For the first time since the dissolution of the USSR, the Russian government actually delivered the sums to science promised in the budget. Also, the number of researchers in Russia began to increase after years of decline. However, the increase turned out to be temporary, due to the economic crisis of 1998. Since then researchers continue to leave. The increase was modest and only in the government research sector and among private enterprises; in higher education and among non-governmental organizations (NGOs), the decline continued. Perhaps most noticeable was that the increase in research personnel in private enterprises—1.2 percent—was higher than in any other area.[22]

At the same time that the government began to fully fund its budgetary obligations to scientific research, it also increased its financing of education. In the budget for 2001, the allocations for education grew by 51 percent over the previous year.[23] Also beneficial to education was the large increase in the budget for the development of telecommunications and information technologies. The federal program entitled "Informatization" became one of the most costly federal programs.[24]

The increase in the number of researchers in the business enterprise sector was an important sign of the beginning of recovery, because up until that time it had been exactly this sector that had experienced the highest outflow of personnel and the most serious retraction. This improvement in enterprises may also be attributed to the growing attention the government gave to the defense sector. President Putin, speaking at the May 2000 meeting of the General Assembly of the Russian Academy of Sciences, announced that support of defense-related research and development was one of the major priorities of his policy toward science and technology. A realization of this policy could be seen in the growing expenditures by the government on defense-related research and development. A noticeable redistribution of resources from the civilian sector of the economy to the defense sector was occurring. As table 2.7 shows, the share of expenditures on defense R&D in percent of total R&D increased from 23.8 in 2000 to 31.6 in 2004.

Besides defense-related science, other priorities announced by the president included the support of fundamental science, the support of young scientists to encourage other young people to enter the field, the development of modern means of telecommunications, and the stimulation of innovation. Putin also gave special attention to the government science foundations (to be discussed in chapter 4). His remarks, however, were somewhat confusing. On the one hand, he said their activities must be "brought under control"; on the other hand, he

Table 2.7. Expenditures on defense-related
R&D out of total expenditures on R&D in Russia

Year	Expenditures on R&D, % GDP	Expenditures on civilian R&D, % GDP	Share of expenditures on defense R&D, in % to total R&D
2000	1.05	0.8	23.8
2002	1.25	0.9	28.0
2004	1.17	0.8	31.6

Sources: *Nauka Rossii v tsifrakh—2002. Statistichesky sbornik.* TsISN, Moscow, 2003, pp. 126–127; *Nauka Rossii v tsifrakh—2004. Statistichesky sbornik.* TsISN, Moscow, 2004, pp. 178–179; *Nauka Rossii v tsifrakh—2005. Statistichesky sbornik.* TsISN, Moscow, 2005, pp. 68, 181.

said that foundations are one of the most effective mechanisms of financing science.[25]

While the central Russian government was increasing its attention to science, the regions of the Russian Federation were doing so as well, and the sources of financing of research and education outside the federal budget began to grow. Several industrial groups gave more attention to education and, to a certain extent, to young scientists. The first such groups to support undergraduates were the Russian Credit Bank, the philanthropic holding foundation INTERROS (supported by Vladimir Potanin), and the oil company YUKOS (later disbanded by the Putin government). As a rule, stipends were given to the best students in higher education in specialties important to supporters of industry and in those regions where the industrial groups are located. Several programs were designed to give financial assistance to the faculties and departments of higher education institutions, with the largest number of students receiving stipends from industrial groups.

The Russian government and the Russian Academy of Sciences devised special programs to support young scientists. Usually the support was given in the form of various types of prizes, grants, and stipends awarded on a competitive basis. However, these measures have not changed the situation very much, since relatively few young people were supported and the support was only temporary.

In the early and mid-nineties science in Russia experienced one of the most calamitous falls in funding and government support in the entire history of science. Only the situation of scientists in Germany and Japan at the end of World War II might be comparable, but Russia had not been engaged in a major war. Furthermore, the science communities of Germany and Japan in 1945 were much smaller than that of the Soviet Union in 1991.

The combination of loss of funding and the subsequent enormous brain drain dealt a grievous blow to Russian science, so serious that its very viability was doubted by many. At the same time, science lost the ideological support that it received under Soviet rule. The old Soviet Union was strongly supportive of science, both financially and ideologically, equating Marxism itself with science, and opposing religion in the name of its "scientific worldview." The disappearance of the Soviet Union brought a rush of new interest in religion and a significant turn against science. The perception of many scientists in post-Soviet Russia that they were no longer appreciated was in many ways a social reality.

The worst part of the crisis lasted about ten years. By 2000 or 2001 the nadir had passed, although many problems remained. After these years attention turned to the questions of the reform of the structure and financing of Russian science and debates about how to make Russia more competitive globally in high technology.

3 Major Directions of Reform in Russian Science

Periodization of Reforms

Government policy on science and technology (S&T) in the post-Soviet period can be divided into three stages: 1991 to 1996, 1997 to 2001, and 2002 to the present. Throughout all these periods the government stated that its goal in science was "reform," but this goal was pursued in ways that were different, often ineffective, and sometimes even contradictory.

The most important changes that have been made in the structure of Russian science in the last fifteen years were the creation of science foundations, especially the Russian Foundation for Basic Research (RFBR, 1992) and, potentially (if it succeeds), the creation of an innovation infrastructure. The "governmentalization" of the Academy of Sciences, which came in 2006, has the potential for being another important change if it is followed by genuine reforms in the Academy, but that has not yet happened. Chapter 4 treats the new foundations, and the new efforts to create an innovation infrastructure are the subject of chapter 5. Finally, the change in the status of the Academy is discussed later in this chapter.

First Stage: 1991–1996

The first stage was closely linked to the policies of B. G. Saltykov as Minister of Science. These were also the years of privatization and "shock therapy" promoted by President Boris Yeltsin. The hope that after enduring a short period of therapeutic pain Russia would make its way to a Western-style free economy with widespread improvement in living conditions was soon dispelled. A small class of entrepreneurs, black marketers, and emerging oligarchs did well in these years, but the mass of the population endured great hardships. People on fixed incomes, such as scientists, teachers, and pensioners, fared especially poorly as inflation eroded their salaries.

The main goal of Minister of Science Saltykov in this period, as reflected in the "Law on Science," was to preserve science in a crisis situation and simultaneously create new elements in the structure of science that might serve as embryos for subsequent reform. It was assumed that these new elements, such as privatized scientific organizations, would compete successfully with the old ones, and as a result the old ones would gradually die out. (The hope for "survival of fittest forms" in the organization of science paralleled, of course, similar hopes for the evolution of economic enterprises in general under the influence of market conditions.) Time would show, however, that these hopes were not justified. Even worse, often the new structures in science would, under the influence of the old ones, gradually begin to resemble the old ones themselves.

Key concepts of science policy in this period were the selective support of high priority programs in science and technology and, indirectly, the scientists involved in them (and not science as a whole); fostering competition and even controversy in science in the struggle for resources; securing the intellectual property rights of researchers; defending the principle of openness and transparency in Russian science; and attempting to include Russian science in the world scientific community.[1] Institutional changes included creating conditions for privatization of some scientific organizations and facilities, promoting an innovation climate leading to greater use of technology in the economy, and fostering financial support for S&T outside the government budget.

During this time legislation was adopted for government science foundations, another foundation for development of technology, and also non-government foundations active in S&T (to be discussed in the following chapters). The first steps were taken toward facilitating foreign philanthropies helping science in Russia, such as tax exemptions and co-financing. The first laws on intellectual property were passed, and the first steps were taken toward bringing research and teaching closer together. All these were new elements in principle for Russia, even if they were incomplete and inadequate.

In order to try to stem brain drain and stimulate science careers, new regulations were passed on additional pay based on earned degrees and qualifications; also, financial supplements were given to young scientists. In addition, those research institutes in applied science that were critical for national development were given special status, indicated by the title "Federal Research Centers" (GNTs). As already noted, all these measures were not enough to improve the situation very much, but they were at least beginnings.

Second Stage: 1997–2001

The second stage was a period of frequent changes in the leadership of science with resulting variety and even chaos in science policy and regulations. The disorder in science was in large part a reflection of confusion in the country as a whole. By early 1996 President Yeltsin had become unpopular because of the distress caused by shock therapy, and it appeared that he would have difficulty being reelected. However, a group of oligarchs gave him strong support and portrayed him as the only realistic alternative to the resurgent Communist Party and its leader Gennady Zyuganov. In the end, Yeltsin managed to be reelected, but confusion and austerity continued right up to the financial collapse of 1998, when Russia defaulted on its debt and the ruble fell dramatically in value.

During these years leadership of the government organizations responsible for science frequently changed. (In the same period prime ministers were repeatedly replaced as well, and the confusion at the top levels of government contributed to the disjointed science policies.) Each new leader of a given bureaucracy announced his own new policy, sometimes even radical in essence. Often the new policy was not carried out. The desire to continue evolutionary reform of Russian science institutions and to preserve scientists was proclaimed, but there was no balance between strategy and tactics in working toward these goals; as a result, tactical moves directed more toward survival than development predominated.

In this period tax privileges were given to research organizations, financial supplements based on degrees and titles were continued, and research institutes were permitted to continue to use income from rents and leases of space in their buildings for the support of S&T activities. Surprisingly, in this period there was almost no new support for young scientists, unless one includes the "Integration" program (to be discussed) begun in the previous period (1996) for the support of "scientific schools." This period of inactivity by the government in the support of science caused a further deterioration of the Russian scientific community. By the year 2000 the number of young scientists in Russia (up to age of thirty) fell by 33 percent from the 1996 level, and the total number of researchers decreased in this period by 20 percent.[2]

In this period there was an effort to improve commercial uses of R&D results. The government began to finance elements of an innovation infrastructure: Innovation Technology Centers were created; there was support for small innovation enterprises and venture investing; and a Russian Venture Innovation Fund was created. At this time the term "National Innovation System (NIS)" began to be used, but the title was still conceived narrowly; it was usually seen as a means of commercializing existing R&D results rather than a reconceptualization of the role of science in the economy.

The end of this period saw an attempt to create agencies to work out a strategic vision on S&T: In 2001, after Vladimir Putin had become the new leader of Russia, a President's Council for Science and High Technologies was created. However, it did not accomplish much of importance. Financial constraints made large initiatives impossible: In 1997 expenditures on civilian science were 0.43 percent of GDP, a proportion that fell to 0.3–0.31 percent by 1999–2001.[3]

Third Stage: 2002–Present

The third stage, continuing to the present time, started in 2002. This recent period coincided with Russia's financial recovery, spurred on by a rise in world oil prices and the increasing competitiveness of Russian goods domestically because of the devaluation of the ruble. The improvements in the financial situation made it possible for the Russian government again to start paying attention to science and to improve scientific salaries and research investment.

This stage began with the active preparation of a large number of conceptual documents directed toward a strategic, long-term S&T policy and other economic initiatives of the government, including industrial policy. The concept of a National Innovation System was somewhat broadened, although it continued to be seen primarily as a system for the practical application of research.

During this period many documents were produced with the goals of preserving the scientific potential of the nation and beginning a systemic restructuring of the S&T establishment. More documents were prepared for supporting and rewarding young scientists and graduate students in this period than in the previous decade.

To restructure the S&T establishment, government accreditation of scientific organizations was repealed, new rules were issued on the commercialization of intellectual property (IP), and also a government inventory of the results of R&D was introduced. A new Russian Patent Law was accepted that introduced a statute easing the definition of rights to intellectual property under government contractual agreements, thus creating a basis for the successful cooperation of government science and private business.

During the previous ten years science policy emphasized preserving the scientific potential of the nation and regulating the results of S&T activity. By 2002, however, it was clear that the old science and the new economy could not be effective together, and as a result a course was taken that aimed toward the systemic reform of science.

Efforts to Reform the Structure of the Public R&D Sector

Starting in 2004 an attempt was made to restructure science as a whole, rather than just particular parts of the scientific establishment. In that year the government adopted a document titled "Conception of Participation of the Russian Federation in the Management of Government R&D Organizations." It proposed reducing the number of government R&D organizations to less than half the current number and shrinking the research staff in the state-owned R&D sector. However, it did not clearly specify the criteria for making these decisions, nor did it establish the responsible bodies and procedures for achieving these dramatic changes. Not surprisingly, little change resulted.

The approach of the "Conception" document in its final form was very old-fashioned and, in many ways, a disappointment. (In its earlier form it was much more radical, reducing the status of the Russian Academy of Sciences, but the Academy managed to defeat this draft.[4]) The Russian Academy of Sciences was given responsibility for fundamental science, Federal Research Centers (many of them the old industrial branch institutes) were given applied research, and the universities were still considered primarily educational institutions. Such a traditional assignment of roles has, in the past, hampered scientific development by separating research, development, and teaching in different institutions. The new conception seemed to continue this separation, although it called for the privatization of many applied-research institutions, including Federal Research Centers.

In the revised "Conception" the Russian Academy of Sciences would retain all its old functions and even acquire new ones. Not only would the Academy preserve its traditional leadership role in fundamental science, but it would also be an ill-defined "coordinator" for government-sponsored applied research. The "self-reforming" assigned to the Academy appeared to be mostly cosmetic, not substantial.

The "Governmentalization" of the Russian Academy of Sciences

On December 5, 2006, President Vladimir Putin signed additions and modifications to the state law titled "On Science and State S&T Policy." This document changed the status and some organizational procedures in the Russian Academy of Sciences. The new provisions were described in some newspapers, Russian and foreign, as the most significant change for the Academy in decades. The leading U.S. journal *Science* ran an article titled "Kremlin Brings Russian Academy of Sciences to Heel," and quoted Nikolai Ponomarev-Stepnoi, a Russian nuclear physicist at the Kurchatov Institute, as saying, "We are being castrated. . . . They are demolishing the academy."[5] The Russian newspaper *Kommersant* published a somewhat similar article, "Russian Academy of Sciences Loses its Independence."[6] And the newspaper *Nezavisimaia gazeta* observed, "Factually, the history of the 'old' Russian Academy of Sciences has ended. Now there begins the epoch of the 'new' Academy—a *de facto* state agency somehow retaining its earlier name."[7]

These are very strong words. But let us look at what the new law actually did. Upon examination we will see that its true significance for the Academy is not at all clear. Under some interpretations, not very much will change. The president of the Academy, Iurii Osipov, supported the changes and said they were "evolutionary," not revolutionary.[8] A vice president of the Academy, Aleksandr Nekipelov, even maintained that the new law "does not worsen but improves the position of the Academy of Sciences."[9] (He was probably referring to the fact that the law gives the Academy several new functions, to be mentioned below.)

Other people maintained that the importance of the new law is not what it does in itself, but what it might permit in the future. According to some of them, the Russian government has at last cleared the way for true reforms of the Academy, even though all admit that these, if they come, are still in the future. The most expert opinion, at the moment, favors the view that not much has actually changed. According to these analysts, although the Academy felt threatened in several of the discussions leading up to the legislation, it has, through amendments to the draft law, managed to preserve itself.

One thing that the law accomplished was much greater clarity of the legal status of the Academy. In the past that status was confused by the fact that, on the one hand, it was often described as a self-governing honorary society, electing its own members and officers, and, on the other hand, it was an enormous network of research institutes totally financed by the Russian government. Which was it, a learned society or a government research establishment? And who actually owned the land and buildings of the Academy, a vast establishment? In Soviet times, when all land and property belonged to the state, questions of ownership were not important, but after privatization occurred in the post-Soviet period in many areas of the economy, what was the status of the property of the Academy? Gorbachev greatly confused this situation by issuing in August 1990 a decree making the Academy fully independent of government influence and able to possess its own property, while remaining totally dependent on the government for its budget. Boris Yeltsin later reversed this decree, but the legal status of the Academy remained uncertain.[10]

The new law removes this uncertainty. It clearly states that the Russian Academy of Sciences (and also the Academies of Medicine and Agriculture) are state institutions. In that sense, the phrase "governmentalization of the Academy" is accurate. The law specifies that the governing charter of the Academy will be approved by the Russian government (the Council of Ministers) and that the president of the Academy will be approved by the president of the Russian Federation. (This assignment of power to the president of Russia fits well with Putin's overall policy of centralization.) However, the draft charter and the nominated president of the Academy will be drawn up in consultation between the government and the Academy. In fact, the president will be elected by the General Assembly of the Academy and then presented to the president of the Russian Federation for confirmation. Will this end up in cases of dramatic conflict? The smart guesses are that it will not, but that instead problems will be ironed out before the nomination is submitted to the Russian president.

Factually, the position of the Academy is not very different from what it earlier was. Both in tsarist times and in Soviet times the government frequently influenced the choice of president of the Academy and, on several occasions, in effect named him or her (there was one female president, Ekaterina Dashkova, a friend of Empress Catherine the Great). Readers of this book will remember that in chapter 1 Academician Vitaly Ginzburg described how the Communist Party in Soviet times also influenced the choice not only of administrative officers but even of the members of the Academy. The government has always been the financer of the Academy and has, through the budget, influenced the salaries of researchers. It is true that the Academy was given considerable leeway in distributing the budget it received from the government, but the new law continues that tradition, saying that the presidium of the Academy will be able to determine the number of scientific researchers, their salaries, and the priorities of research (within the overall budget assigned by the government).

The new legal clarity does make it possible, should the Russian government wish to do so, to make further reforms. Theoretically, it could force through a transfer of many or even all institutes of the academy to universities and industry, using its control over the budget, charter, and leaders of the Academy as levers for this transformation. It might even set up the goal of turning the Academy into a normal (in international terms) learned society, without administrative responsibilities for research. But accomplishing these truly radical reforms would involve a direct confrontation with the most distinguished scientists in the nation, a battle of the sort in which, if the past is any guide, the leaders of the Academy have proven very clever at retaining their most cherished authority and privileges.

One of the most interesting features of the new law is that it permits the Academy to create educational institutions and "innovation enterprises." Much depends on how this provision is interpreted. Does it mean that the Academy can create "research universities" that will compete with the traditional state universities under the Ministry of Education and Science? Does it mean that the Academy can run profit-making innovative industries and become a force in the economy? Much remains to be learned. If such developments occurred on a large scale, the Russian Academy of Sciences might become even more influential than it currently is. One thing is clear: it is premature to pronounce the end of the era of Academy dominance of fundamental research in Russia.[11]

Changes in Government Financing

In Russia, as noted earlier, over 70 percent of all R&D is performed in government organizations.[12] Of the total of R&D organizations 40 percent receive guaranteed government financing based only on the number of researchers in their organizations. Such a budget formula does not stimulate productivity or efficiency in research. At the same time, the government indicated that it would like to support only a restricted number of priorities in research, those especially important to the national economy. Much effort has been made to accomplish this change, usually by calls for reform of the budget process, emphasis on certain priorities, and distribution of funds on a competitive basis, but so far the results have been unimpressive.[13]

A new budget classification, introduced in 2005, divided expenditures on science into ten functional categories (such as "general government problems," "national economic problems," "health," "education," "national defense," etc.). Previously there had been a separate section of the budget devoted to "Fundamental Research and Assistance to S&T Progress," which clearly stated the amounts of money given to fundamental and applied research for non-defense purposes. The new classification complicated analysis of appropriations for non-defense science; now expenditures on fundamental research were concentrated under the category "general government problems," but applied research was scattered throughout all ten categories. In several of these categories, non-defense and defense purposes seem to be mixed.

Applied research is realized primarily through what are called Federal Goal-Oriented Programs. The most important of these seems to be R&D on Priority Directions of Science and Technology for the period 2002–2006 (PDST).[14] In

2005 this program was significantly strengthened by its union with a series of other programs (for example, the one called Integration of Research and Education). As a result, the budget of this program (PDST) is now the second-largest for R&D, after the Federal Space Program.

In retrospect we can see that this PDST program was inherited from the Soviet period. A serious evaluation of its results has never been carried out. It is true that there have been sketchy evaluations of proposals and the status of the research leaders and their teams, but such indicators as quantity and quality of publications, citation analysis, number and size of former grants, and success in gaining industrial contracts have not been considered. As a result the scientists in this program who in the West would be called "principal investigators" have been the same people again and again; new funds, projects, and people for stimulating innovation have very rarely appeared. Only very recently (2005), with the formation of a new, strengthened PDST program, have some of these problems started to be addressed. An effort is currently under way to strengthen competition and at the same time the amount and continuity of financing.

One would think that the strengthening of such projects would have a positive effect. However, it seems that the situation may actually be worsening. The increase in the size of the program at a time when the overall budget for science is growing only at a modest rate means that the number of project leaders within the PDST program is decreasing. In the absence of additional funds from industry or elsewhere, the project leaders in the PDST program have resorted to increased lobbying and have also used corrupt means to gain support.[15] Most evaluations have shown that the identities of the project leaders were known in most cases even before the competition began. The lists of winners in the competitions show that there is a rather permanent narrow circle of successful organizations and leaders. Beginning in 2004 the government tried to make evaluations based on results rather than superficial evaluations centering on seniority and past associations, but the criteria for doing so remain unclear.

At the same time the Ministry of Education and Science has tried to formulate more precise goals in the sphere of science and innovation, focusing on "the creation of conditions for the development and effective use of S&T potential" and "the creation of conditions for innovation activity." Such formulations are, unfortunately, based more on process ("creation of conditions") than on desired results. Measurement of actual performance is still largely missing. Complicating the situation is the fact that most existing criteria for attainment of goals are designed to obtain additional funds, not on ability to achieve actual results. Not enough attention has been paid to the experience of research managers in other countries who have long wrestled with these problems and have developed appropriate criteria. They have learned that quantitative criteria often fail to yield successful research results. Therefore, they have usually relied on peer-review systems that include qualitative evaluations. So far such systems have not been taken into account by Russian government officials developing new budgets.

The Problem of Brain Drain

Brain drain continues to be a problem despite some efforts to diminish it. Emigration proceeds now at a significantly lower level than in the nineties, but the seriousness of this problem is still evident among young scientists and recent university graduates.

In the last five to seven years the emigration of youth has become the main component in brain drain, in contrast to earlier years when many middle-aged and senior scientists left Russia. A survey conducted in 2003 among recent graduates of Moscow University[16] showed that 10 percent of the biologists, 11 percent of the physicists, and 13 percent of the chemists already received invitations from abroad several months before graduation—striking statistics since these invitations are based not on actual research accomplishments but on assessments of potential. A similar situation appeared in the results of a survey of young scientists working in research organizations in Moscow: 44 percent of the respondents hoped to go abroad at least temporarily, and 7 percent said they hoped to remain there permanently.[17]

The existing large Russian scientific diaspora plays an important role in stimulating young scientists to leave. Scientists abroad who earlier lived in Russia and are now successfully working in foreign laboratories help their former countrymen to receive grants and make R&D contracts. In the process they monitor and select the best undergraduates and graduate students in Russia and often invite them to come abroad to continue their work. The scale of such contacts and invitations is not known and would be very difficult to define, but anecdotal evidence indicates that it is widespread. Scientists from practically all the strongest Russian research institutions have emigrated abroad, and many of them keep contacts with their "mother" organizations. Foreign countries often facilitate such outflows. In the UK, Germany, Canada, and New Zealand immigration laws ease the inflow of highly qualified personnel to post-doctoral positions.[18]

At the same time, in several scientific centers in Russia the brain drain is diminishing. As a rule this is happening in research institutes specializing in rapidly growing research themes such as biophysics, molecular biology, biotechnology, new materials, etc. But to say that in whole fields of research brain drain has stopped would be inaccurate. In a few research organizations (examples are the Institute of Molecular Biology of the Russian Academy of Sciences, the Institute of Catalysis of the Siberian Division of the Academy, and the Institute of Theoretical and Experimental Physics) many measures have been taken to keep young people in science; there is stable financing by government and foreign sources, good contacts are maintained with foreign research centers, new equipment has been obtained, and young people have been given means to advance their professional careers. These measures show that there are ways to reduce the outflow of young scientists.

Brain drain is, of course, a world-wide issue, not a uniquely Russian problem. In some countries in the Caribbean and Central America over 50 percent of college graduates leave their home countries.[19] The problem afflicts areas such as Canada and Western Europe as well as less-developed nations. The University of Toronto in Canada recently instituted a number of prestigious professorships designed to prevent its faculty, especially in fields like nano-technology, from migrating south to the United States.[20] Some laboratories in the United States have been described as "miniature European Unions."[21] In a globalized world, scientists will inevitably migrate to the spots where they think they have the best opportunities. The United States itself is vulnerable to the same phenomenon.[22] The most that any country, including Russia, can hope for is achieving a situation in which the flow is at least as much inward as outward.

Government Policy on Scientific Personnel

The number of legal measures taken by the government regarding scientific personnel exceeds all others relating to science. Most of them have been pay raises, prizes to scientists and science administrators, support for "leading scientific schools," and increases for stipends and grants to young scientists and graduate students. However, experience and the results of surveys show that these measures alone cannot lead to a solution of the brain drain problem.

In 2003 the Ministry of Industry, Science, and Technology tried to address the personnel problem in several conceptual documents envisioning developments in 2004–2009. Its program titled Scientific Personnel of the Russian Federation was dedicated to keeping the most productive scientific researchers in Russia and meeting the needs of young scientists. The proposed program included various forms of financial incentives, grants, bonuses, support of "scientific schools," and a system that would provide home mortgages for scientists and engineers.

So far, however, the only concrete result of the proposed changes was an increase in pay to young scientists holding the degrees of "kandidat" or "doktor" who were winners in a special competition, and also an increase in the size of prizes given by the government and the president. Up to the present the personnel problem has been addressed primarily by these traditional measures: selective and temporary increases in pay for a few categories of employees.

The Russian government has not appropriately addressed the general problem of eliminating or reducing the negative factors that drive scientists either abroad or out of their professions. Little has been done on the government level that might help a reverse flow of scientists from other countries into Russia and therefore compensate for the outflow of recent years.

Officials in the Russian government are not of one mind on the issue of brain drain. Some government officials do not see the departure of scientists as a special problem, viewing it only as a part of normal migration. There are others, however, who believe that the country is suffering great financial and intellectual losses from brain drain and who would consider limiting or prohibiting the emigration of scientists. The problem is one, therefore, of considerable political significance on which there is no agreement.

A separate political issue concerning scientific personnel is the fact that in recent years several Russian scientists have been jailed on charges of divulging state secrets, raising fears among human rights organizations abroad that Russian security organs have returned to some Soviet practices. These scientists include Aleksandr Nikitin, a nuclear reactor engineer who complained about environmental abuses in the Northern Fleet; Igor Sutiagin, a nuclear scientist at the Institute for USA and Canada Studies, who was accused of passing on sensitive information about Russia's weapon systems to U.S. military officers; and Valentin Danilov from Krasnoiarsk State Technical University, who was accused of providing classified information to China. Each of these cases is different, but a common element is that Western organizations have protested that the Russian government has regarded as "classified" information that was actually publicly available, and that the arrest and trial procedures have sometimes been closed and not in accord with international standards.[23]

So far measures taken on the level of the government are only local and fragmentary and do not take into account the general context of science in Russia. The personnel problems of science could be solved by a simultaneous improvement of the general situation of science and the application of specialized measures for supporting scientists. General measures could include stimulating industry to invest in R&D by changing tax laws; developing indirect measures for regulating and encouraging government–private partnerships; funding equipment and facilities for R&D in universities where obsolescent equipment is one of the main factors causing students to leave science; encouraging mobility of researchers between the government and private sectors of science; and supporting an infrastructure for innovation.

Valuable special measures would be introducing post-doctoral positions in universities and other scientific organizations, and supporting the social welfare of scientists in retirement. The general fate of retirees in Russia is dire; retired scientists are usually unable to do even part-time work because they do not have access to their laboratories. Some U.S. universities have tried to address the professional dimension of this problem by allowing senior faculty to shift to half-time work and keep their access to laboratories and libraries. Such policies would help the retired scientists while making room for the career growth of young scientists and therefore help keep them in science. However, in Russia, where R&D organizations and universities are usually federally owned, the strict rules on staff members make such policies difficult.

This short overview of the major directions of reform in Russian science shows that during the last fifteen years the Russian government, in several different stages, has made attempts to formulate goals and tasks to develop science and innovation. The government has tried to find ways to form stable linkages between participants of what government administrators call the National Innovation System. The government has asserted tighter control over the Russian Academy of Sciences with the hope of enacting further reforms in the Academy, as yet unrealized. It has raised, several times, the issues of structural reform, changes in financing, and personnel problems, including brain drain. Many different government documents on science and innovation have been formulated, new laws have been passed, and new organizational entities, such as government science foundations, have been created. Foreign experience has been studied and sometimes adapted.

However, the results of these efforts at reform are often less significant than hoped and expected. So far the government has expressed positive intentions and formulated documents embodying these intentions, but it has not constructed a careful plan of action that would bring them to reality. This is true for both organizational and financial reform. Furthermore, it has met resistance from older structures and organizations that has often blunted, if not defeated, its original intentions.

Most of the initiatives of the government are of a fragmentary nature. Priorities of importance are often lacking, the efforts are poorly coordinated, and their results are inadequately evaluated. Poor planning has hampered the policies on scientific personnel and on the integration of research and education (to be further discussed in chapter 7). Some of the initiatives have been inconsistent with earlier ones and have demonstrated a lack of understanding of the role of science in the entire economy. Often science has been conceived only in terms of certain applications rather than in terms of how it can be integrated thoroughly into the economy in a way that diverse and unexpected applications will naturally arise and be used. In short, a broad interpretation of science and innovation is still lacking, and concrete measures to implement even those fragmentary and somewhat narrow reforms that have been proposed have usually not been taken.

4 Foundations: A Novelty in Russian Science

The fall of the Soviet Union led to important organizational changes, not least in the funding of what had been internationally recognized as one of its premier activities, its scientific research. The new Russian government, observing what it considered to be the success of research in the United States supported by the National Science Foundation and the National Endowment for the Humanities, broke with the exclusive Soviet tradition of block funding of scientific institutions and created science foundations supported by the federal budget. These new institutions relied on applications from individuals or groups of scientists who wished to pursue research on specific problems. At the same time, foreign foundations also became active in Russia, and they pursued their own methods and priorities, many of which were different from those customary in Russia. And finally, particularly in the most recent years, a few domestic Russian non-governmental foundations appeared, financed by newly-rich entrepreneurs, people who also had their own priorities and methods. Thus the funding of research in Russia has become immensely more complex than in earlier Soviet years.

Foundations as funders of scientific research have had a great impact on the Russian scientific community, as will be discussed below. At the same time they have raised expectations in ways that can probably not be fulfilled, and therefore have increased frustration.

The creation of science foundations by the Russian government was not only its attempt to modernize its methods of funding science; it was, at least in the beginning, also an effort to ease a deep financial problem. The new Russian government had much less money at its disposal for scientific research than the old Soviet one, and creation of a method of funding based on competition was one way to handle this problem. Thus the foundations were not seen by scientists, especially in the early and mid-nineties, as an entirely positive step. They were also reminders that acquiring money for research was going to be much more difficult than in the past. In that sense, they were bad news for quite a few Russian scientists, and even the majority of scientists were initially skeptical. Gradually, however, these attitudes began to change, as described below.

Fitting these new organizations into the existing set of scientific institutions and traditions in Russia has not been easy; Russian scientists, science administrators, and politicians have long debated the best way to accomplish this task. The discussions that have occurred in Russia over the place and function of science foundations are of interest not only to Russians but also to people in other countries, including the United States. In fact, in their attempts to conceptualize and legalize the role and functions of science foundations the Russian have posed issues with a clarity that can be enlightening to scientists and politicians in all countries. Just how do grant-giving self-governing foundations fit with government policies? What are the rights of governments if they supply the funds these foundations use? Who should own, or have the rights to, the ideas and innovations resulting from research financed by the foundations? To what degree should governments regulate foundations, and what difference does it make if some foundations are government-financed and some are not? What difference does it make that some foundations are foreign and some are domestic? To what extent should foreign foundations be permitted to determine research topics and research priorities in another country? If in Russia—with its traditions of centralized state control of politics and economy—these questions took on an unusual urgency, they nonetheless have a relevance for all countries. Americans tend to be a little blind to the significance of such issues for several reasons: they are so accustomed to the activities of foundations that they tend to take them for granted, and they have never confronted the possibility that a foreign foundation might come to the United States and disburse, according to its own priorities and methods, sums of money so large that they competed with domestic sources. But that is exactly what happened in Russia in the early and mid-nineties, as will be described in chapters 6, 7, and 8. First, however, we will look at Russian governmental foundations, and only later consider foreign and domestic non-governmental ones.

At the time of the collapse of the Soviet Union, no science foundations of any type existed. Research institutions were funded by the government on the basis of past budgets and numbers of researchers employed at a given institution. There was no system of peer review in science, nor was there a system for financing specific teams of researchers. The absence of peer review meant that the opinions of the scientific community itself were not very important in influencing the priorities and themes of research. It is true that top administrators in science, people who had earlier often been research scientists themselves—such as the members of the presidium of the Academy of Sciences—had some influence on such questions, but this system was based on "top-down" principles. Local researchers, particularly junior ones, had little influence.

The creation in 1992 of the first government science foundation—the Russian Foundation for Basic Research (RFBR)—brought a profound change in principle to the support of scientific research in Russia. It must be emphasized that this change was more in principle than in fact because, as we will see, the budgetary funds available to the RFBR were very modest. Much of Russian science continued to be funded as before, from the top down. Nonetheless, a significant change occurred with the creation of the RFBR: this act established new principles of financing and a new mechanism of distribution.

During the organization of the RFBR Russian officials studied foreign models of science foundations, especially the National Science Foundation in the

United States.[1] In addition they visited the National Institutes of Health. The Russians were also interested in the Deutsche Forschungsgemeinschaft, or DFG, in Germany.[2] The most important feature the Russians took from these foreign models was the peer-review process. However, these foreign foundations were not duplicated, but instead used as examples that could be modified according to local customs and preferences. In the NSF, for example, program officers are usually only temporary administrators, on leave from U.S. universities. Such rotation is designed to contribute to the flow of fresh ideas to NSF from the academic perspective and to demonstrate to the visitors how NSF works. Partly for economic reasons (travel and housing expenses) and partly for legal ones (underdevelopment of a contract system for hiring staff), the RFBR has relied primarily on permanent personnel, usually from Moscow. Not surprisingly, this difference has led to criticisms, to be discussed below.

Nonetheless, the RFBR differed dramatically with the Soviet administrative–command system that preceded it. It brought both a new conception of how science should be governed and a new method for the financing and management of those research projects that are approved.

The new conception upon which a science foundation is based is that researchers should have much more freedom than was the practice in the Soviet Union. Each researcher or research team is free to submit an application independently and to hope to receive government funding for a given research project as designed by local scientists. This conception differs in principle from the Soviet system of the direction of research by leading administrators.

In this new system the initiative of researchers is encouraged. Permission to submit proposals either is not needed, or, more accurately, is routinely and easily given (something especially important for fundamental science). This initiative transforms the "cutting edge" of research, forming it from below rather than from above. By transferring emphasis from the institution and its top administration to the local researchers, interdisciplinary research and cooperation among researchers from different institutions is encouraged. The research problems in science become more important than the organizational affiliations of the scientists.

Although foundations brought much more freedom to Russian scientists in choosing their topics of research than the old Soviet system, of course this freedom also has its constraints and limits. Foundations often have priorities that are reflected in the amounts of money available in a given area. Proposals must fit within those priorities. Nonetheless, compared to the previous Soviet method of funding, the foundations greatly enlarged the area of initiative for individual scientists.

Foundations also bring with them a new method of project selection and subsequent administration. Projects are chosen by peer review administered by the foundation. Peer reviewers are, at least in the best of circumstances, highly qualified independent scientists who are able to compare the proposals and the credentials of the applicants to the best in the field.

The funding of research by foundations usually operates administratively according to the principle of transparency, with accountability and control over the expenditure of funds. The foundation does not normally interfere in the scientific research itself, but it requires records and receipts for expenditures, and it can ensure that the original budget is either observed, or changed only within permissible limits. To be sure, the degree of transparency and freedom granted by foundations is relative, and abuses do occur, but financing in this new way is more exact and less prone to corruption than the block-funding method common in Soviet times.

Essential to the foundation system of financing is the concept of the grant, a term that did not exist in the old Soviet Union. The idea of a grant has a history of its own in the Western countries where it arose. The terms "land-grant" and "grant-in-aid" go back centuries, but the concept of a grant given to a researcher, or group of researchers, by a philanthropy did not come into currency until after the creation of large foundations in the United States in the early twentieth century. By the late 1920s and early 1930s the Carnegie Corporation (created in 1911) was regularly referring in its annual reports to "grants" and "grant-making committees."[3] These terms gradually spread in the United States, where they achieved legal status and wide circulation. In the natural sciences the notion of the grant became particularly common in the United States after the creation in 1950 of the National Science Foundation (NSF), which financed fundamental research in U.S. universities on the basis of grants. Grants, in contrast to contracts, are used frequently for the support of basic research. In such cases the results are usually indefinite and do not bring direct assistance or profit, at least not in the near term. The Russians in the 1990s gradually assimilated these terms and practices from Western experience.

Since in the Russian language the term "science" (*nauka*) includes both the natural and social sciences, at first it was thought that one foundation—the Russian Foundation for Basic Research, or RFBR—could handle all knowledge. However, it soon became clear that the natural scientists who dominated the RFBR favored their own fields, to the detriment of the social sciences. Furthermore, the social sciences and humanities in Russia still suffered greatly from the years of ideological deformation in Soviet times and were in need of special attention. In the early and mid-nineties some fields—such as philosophy, political science, and history—were still crippled by the effects of these earlier ideological constraints. Moreover, scholars in the social sciences and humanities did not have as many opportunities for ministerial and industrial contracts as did those in the natural sciences. For a number of different reasons, then, the social sciences and humanities needed the type of special attention that only a separate foundation serving them could give.

Russian administrators of research also noticed that in many other countries the social sciences have their own funding agencies and research centers—such as the Maison des Sciences de l'Homme in France and the Social Sciences Research Council and the National Endowment for the Humanities in the United

States. Therefore, in 1994—two years after the creation of the RFBR—a separate Russian Foundation for the Humanities (RFH) was established. Although it supported different fields than the RFBR, it operated on the basis of similar rules, mechanisms, and regulations. However, among the recipients of grants from the RFH were more individuals and small scholarly groups than in the case of the RFBR, reflecting common patterns of research in the social sciences and humanities. Soon proportionately more social scientists and humanists participated in the competitions of the RFH than natural scientists in the RFBR, a result, no doubt, of the fact that they had fewer other potential sources of support than the natural scientists.

When the activities and methods of operations of the new government foundations in Russia—the RFBR and the RFH—are compared to Soviet practices, one sees the effects of an open market economy on science. Just as in the economy, competition and individualism become important characteristics. The general principles of science funding by the foundations emphasize individual initiative ("bottom-up"), independence, personal accountability, and peer review. Bureaucracies do not usually interfere once projects are under way except in financial accounting. Individuals are expected to take responsibility for their research projects and are obligated to give a full report of results after the project is completed.

The current charters of the RFBR and the RFH state the general principles of their operation, many of them borrowed from foreign models. Although grants are administered by institutions, they are predominantly under the control of the scientists conducting the research (called principal investigators in the United States). Grants are, in principle, made without consideration of age, scholarly title, academic degree, or position in scholarly organizations, a policy especially important in Russia, where favoritism among senior scientists is extremely common. The most important criteria are the scholarly quality of the proposal and the potential of the group or applicant to carry it out successfully.

Despite the importance of the individual researcher in the new rules, however, the institution through which the researcher applies also benefits in the form of overhead—15 to 20 percent of the grant itself. In this way those institutions in which there are many grant recipients automatically receive much more support than the others. Thus the new rules favor both strong individuals and strong institutions over weaker ones, as befits a market economy.

Selection of winners in this competition is based on a multi-stage peer review of submitted proposals. Although, as we will see, this selection process is not entirely free from institutional and personal pressures, it is a great improvement over the earlier Soviet system.

The money given for research by the new foundations comes in the form of grants, not loans. The researchers promise to conduct their research in the direction indicated, and they also promise to make their results freely available to all, but if they do not fulfill these promises the only punishment (assuming

Table 4.1. Financing of RFBR from the federal budget

Year	Budget, mln. rubles, in 2002 prices	% of growth from the previous year
2002	1,734.5	—
2003	1,753.4	plus 1.1%
2004	1,900.2	plus 8.4%
2005	2,417.7	plus 27.2%
2006	2,836.6	plus 17.3%

Source: Ministry of Finance and IET calculations.

there was no financial fraud) is the discrediting of their reputations, and the resulting small probability that they will receive another grant in subsequent competitions.

Emphases of the New Foundations

About 60 percent of the money distributed by the RFBR goes to the support of research projects proposed by researchers (including groups up to ten people in number). The RFH distributes more than 50 percent of its money in a similar way. Both foundations give the rest of their money to a variety of other activities, including travel, expeditions, publication costs, electronic libraries, youth programs, and public outreach. (The reason the RFH gives about 10 percent less to independently initiated research projects than the RFBR is because publishing books is a greater need in the social sciences than in the natural sciences, and therefore more money is set aside for publication costs. In the natural sciences articles are often more important than books.)

The budgets of the RFBR and the RFH are, according to current legislation, a fixed share of the total government expenditures on civilian science. That legislation stipulates that the foundations should receive 7 percent of the total government expenditures on civilian science.[4] The inadequacy of this share is illustrated by noting that in the United States the National Science Foundation, which served as a model for the RFBR, traditionally receives about 20 percent of total federal government expenditures on basic research at academic institutions. But the situation was even worse than these statistics indicate, since often the RFBR did not, before 2002, actually receive the funds stipulated in the legislation. In the first six months of 1998, for example, the RFBR received less than one-sixth of its designated budget.[5] The RFBR budget vacillated wildly before 2002. The years 1999, 2000, and 2001 were rather good ones, since the budget came closer to being fulfilled. Starting in 2002 the RFBR and the RFH received their budgets in full. (See tables 4.1 and 4.2 for recent RFBR budgets, and for distribution of funds within these budgets.)

Table 4.2. Distribution of RFBR budget by programs

Supported activity	2004	2005	2006
Research projects (including grants for organization of conferences and travel grants)	83.5	63.2	59.5
Commercially oriented fundamental research projects	4.7	10.7	9.1
Fundamental basis for engineering sciences	—	4.3	6.3
Equipment grants	—	8.4	6.3
Electronic libraries	5.8	5.4	4.9
Regional competitions and competitions with FSU countries	2.7	2.8	5.6
International competitions	—	2.2	2.8
Joint competitions with Russian ministries and agencies	—	—	2.8
Overhead	3.3	3.0	2.7
Total	100%	100%	100%

Source: *Poisk* for respective years, as reported by RFBR for those years.

The comparison given in the previous paragraph of the shares of NSF and the RFBR in financing fundamental science understates the amount of money going to research in the United States through a peer-review system, since, in addition to NSF, there are many other channels of government science funds disbursed on the basis of peer review. These sources include the National Institutes of Health, the Department of Defense, the Department of Agriculture, the Department of Energy, and the National Oceanic and Atmospheric Agency.

Despite the smallness and irregularity of the amounts of money disbursed by the RFBR in Russia, the grants are nonetheless helpful. The grant funds are usually spent on salaries (up to 50 percent of the total grant), but also for computer technology and instrumentation, travel costs to conferences, and supplies. According to surveys, scientists see the grants as especially important for instruments and equipment, although the money given is not usually large enough for major purchases.

Over the years the RFBR and RFH have increased the number of initiatives they undertake, with new programs to support libraries, young researchers, innovation projects, telecommunications, science publishing, and purchase of equipment. The most dramatic needs vary with time. Thus, in the early and mid-nineties, when scientific publication was financially most difficult, the foundations helped academic publishing. Beginning in 1995 the foundations tried to alleviate the severe shortage of equipment and instrumentation in science both by providing more funds for equipment and by establishing centers for sharing equipment. In 1999, alarmed by the aging of the Russian scientific community, the RFBR announced a special program to attract and keep youth in science. By varying its initiatives the foundations are obviously trying to meet the needs of Russian scientists, but always, in the opinion of the scientists, inadequately. The distance between the needs and the resources always remains large.

In 1997 the foundations began an attempt to attract additional support for science by sponsoring regional competitions in which local governments are challenged to match, or supplement, federal funds given to researchers. The potential for combining sources of funds in this way seems significant, since local budgets and industries have improved in recent years. Traditionally the portion of budgets of regional governments given to science has been very small, almost negligible. According to 2003 statistics the situation improved slightly; in that year the share of regional budgets given to science was on average 0.1 percent to 0.2 percent, and the maximum was 0.7 percent. The legal basis for the joint financing of research by federal and local budgets has, however, remained disputed.[6]

During the last ten years the RFBR has evaluated in total more than 130,000 proposals. More than 30,000 research and publishing projects have received support, projects in which more 150,000 scientists from 1,500 different institutions have participated or are currently participating. Table 4.3 lists the different programs supported by the RFBR in the period 1993–2005.[7]

The RFH in the period 1994–2000 selected on a competitive basis and then supported almost 17,000 projects, more than a thousand scientific meetings in Russia, and more than a thousand trips by Russian scientists to foreign scientific conferences. It also supported more than 500 scientific expeditions and other field investigations, and more than 700 telecommunication and information system projects. In addition it subsidized the publication of more than 3,000 scientific books and helped distribute 300,000 copies of books to libraries. Because of the competitive nature of selection and the use of peer-review methods of evaluation, it is fairly safe to assume that the effects of the RFH during these years were positive for research in the social sciences and humanities, promoting new scholarship and new themes; one should readily admit, however, that we have no good way to measure waste in research. The distinction between "the new" and "the trendy" is sometimes difficult to draw. Research in all countries is an inherently wasteful enterprise, with only a few of its products of world-class significance, but no one has yet devised a better way to improve quality than by competitive selection evaluated by recognized scholars, and the Russians were clearly moving in this direction.

The RFBR and the RFH supported about 40 percent of the proposals they received, which shows that applicants had a reasonable chance of success. Although foundation officials (in all countries) sometimes say that the optimal situation is when a peer-reviewed competition supports about 20 percent to 35 percent of applications, obviously "optimality" here depends heavily on the average quality of the proposals.[8] Within a given scientific community, morale is helped by the belief among its members that the competition for money is not so stiff that the chances for success are negligible. The RFBR and the RFH were obviously sustaining morale in that sense, but the small size of the awards

Table 4.3. Programs carried out by the RFBR in the period 1993–2005

Competition type	1993	1994	1995	1996	1997	1998	1999	2000	2001	2002	2003	2004	2005
Research projects	•	•	•	•	•	•	•	•	•	•	•	•	•
Information, computer, telecommunication projects for R&D		•	•	•	•	•	•	•	•	•	•	•	•
Publications	•	•	•	•	•	•	•	•	•	•	•	•	•
Equipment grants		•	•									•	•
Support of journals		•	•										
Centers for shared use of equipment		•	•	•	•	•	•	•	•	•			
Expeditions	•	•											
Scientific conferences in Russia	•	•	•		•	•	•	•	•	•	•	•	•
Scientific conferences abroad	•	•	•			•	•	•	•	•	•	•	•
Regional competitions						•	•	•	•	•	•	•	•
International competitions			•	•	•	•	•	•	•	•	•	•	•
Library support					•	•	•	•	•	•	•	•	•
Electronic libraries						•	•	•	•	•	•	•	•
Software for supercomputers									•	•			
Science popularization competitions						•			•	•	•	•	
Analytical overviews of competitions									•		•		
Grants to young researchers													•

Source: V. Konov, N. Lialiushko, A. Blinov, "RFFI: 14 let sluzheniia Rossiiskoi nauke," *Obshchestvo, osnovannoe na znaniiakh: novye vyzovy nauke i ucheny·m. Materialy mezh-dunarodnoi konferentsii* ("RFFI: 14 years of service to Russian science," *A knowledge-based society: New challenges to science and scientists. Materials of an international conference*), Feniks (Phoenix), Moscow, 2006, p. 147.

often diminished the positive effects. The average size of a grant was $9,000 a year for a group of up to ten people. The Russian foundations had made the choice to "support more with less" rather than "fewer with more."

The foundations were not, and were not intended to be, a full replacement of block funding. They did not take on the responsibility, for example, of adequately renewing scientific equipment, or supplying all the services scientific researchers need, such as information technology and library resources. An important issue in Russia in this formative period has been finding the correct balance between traditional state financing from above and new grant opportunities obtained from below. Throughout the period traditional budget support by the federal government, based on existing scientific organizations, has continued, even if at a very inadequate level. In these conditions the foundations have sought, also inadequately, to close several gaps and to alleviate the most urgent needs in science. The members of the scientific community have gradually become more aware of this role being played by the foundations, and have increasingly given them appropriate notice.

In sum, the foundations have become an integral component of the science establishment of Russia, recognized by the researchers themselves, their administrators, and the government. Receiving a grant from the RFBR or the RFH is now considered one of the important criteria for evaluating the quality of researchers in a given institution, and the total number of such grants received is a criterion for evaluating the institution as a whole. Such success is taken into account in the certification of scientific workers and laboratories and in governmental accreditation of research institutions. Furthermore, several foreign foundations now consider previous success in RFBR and RFH competitions when they evaluate current applicants.

By their activities the foundations have helped reveal the number of actively working scientists in Russia, as well as their institutional locations and topics of work. All active scientists may not have applied to the foundations, but the belief is widespread that most such scientists have indeed done so. Each year 20,000–25,000 researchers apply to the RFBR, and 10,000–12,000 similarly apply to the RFH. The total number of researchers working in the three main academies (the Russian Academy of Sciences, the Russian Academy of Medical Sciences, and the Russian Academy of Agricultural Sciences)—locations where most fundamental research is carried out—is 83,000, according to statistics for 2002.[9] About half of all grants given by the foundations have gone to researchers in these three academies.[10] Basing themselves on these statistics, some observers have concluded that only between one-third and one-half of all scientists in the academic sector are doing active research work. To be sure, this observation is based on the assumption that most active researchers apply to the foundations, an assumption perhaps not correct. But even if one allows for a small number of creative researchers who never apply for financial assistance, the impression is still strong that the current Russian research establishment contains much "dead wood," maybe even as high as 50 percent. Such an impression has strengthened the hands of administrators who wish to cull the Russian science establishment.[11]

It is possible that the disturbing implications of these statistics explain the continuing controversies over the science foundations. Some scientists and science administrators wish to reduce the independence of the foundations, to place them under greater control by some kind of government department more willing to finance science according to political, rather than scientific, priorities. In addition to supporting research itself, the grants given by the RFBR and the RFH provide recognition and status among scientists in Russia; the less recognized ones are obviously displeased by this revelation of their lesser standing. Even some science administrators are unhappy, since their previous powers to rule their organizations, choosing favorites, have been diminished. Not only scientific success but also organizational influence increasingly comes now from receiving grants from the foundations. These are unsettling developments in Russian science, causing continuing conflict and debate.

Opinions in the Russian Scientific Community about the Foundations[12]

As the statistics below show, predominant attitudes among scientists toward the foundations have changed in the years since their activities began. But in assessing these attitudes one should make a distinction between opinions about a grant system in principle, and opinions about the specific Russian science foundations currently promoting such a grant system.

The idea of financing science by grants has found broad support among scientists, and the portion of those who give positive evaluations has steadily grown with time. In 1993–1994, according to surveys, competitive grant-financing was positively evaluated by 75 percent of scientists; just a year or two later, in 1995, that number had grown to 86 percent.[13] In the summer and fall of 2003, in a postal survey of scientific workers in ten regions of Russia conducted by one of the authors, the grant system of financing was evaluated positively by 92.9 percent of male researchers and 80.3 percent of female researchers. This survey covered all foundations, Russian and foreign. (The gender difference triggers questions about the role of women in Russian science and possible gender discrimination, a topic to be discussed later.) By the early years of the twenty-first century many Russian scientists considered the very posing of questions about the necessity of grant financing to be absurd. They had become convinced of the need for such financing, even if they often had questions and criticisms about the ways it was being accomplished in Russia.

Those who welcomed the grant system noted that in the current situation grants are nearly the only possibility for researchers to survive and work, to make trips to conferences, to buy equipment, and to have access to the internet. Grants from foreign foundations, especially, are often large enough to cover many research costs, including salaries. Not surprisingly, grants supplementing the minimum institutional wage are often seen by scientists as necessities (failing a grant, a second job that detracts from research time becomes necessary). The grants help researchers, then, in two different ways: they help pay the actual costs of research, and they may eliminate the need for researchers to take on a second "outside" job to supplement family income.

Researchers responding to surveys also noted that grants give a moral stimulus to work, increase discipline, and encourage more focused goals. They also mentioned that grants are very important to young scientists, not just because grants give material support, but also because they encourage independence in scientific work, strengthening confidence. Grants help young researchers emerge from their earlier dependence on senior mentors.

Yet despite these positive comments about the grant system in principle, the same surveys show that Russian scientists are often less pleased with the specific Russian scientific foundations currently providing grants. Their attitudes toward these foundations are often neutral or even critical. One of the chief complaints is that the foundations favor certain scientists in Moscow and St. Petersburg; consequently, negative evaluations grow as one moves away from these cities. A common criticism is that cliques of scientists that resist outsiders have formed around the foundations in the two major cities.

The argument that the foundations are under the control of cliques in Moscow and St. Petersburg is a serious one. Even the former director of the RFH, Evgenii Semenov, recognized that "those organizations whose representatives work in one capacity or another in the foundation received a large number of grants."[14] Similar criticisms have been made of the RFBR. One of the peer reviewers for that foundation observed that the leaders of well-known science departments and directors of institutes are "constantly (or almost constantly) to be found in the lists of grant-recipients."[15] These statements about the existence of "clans" are difficult to confirm or refute, since the foundations do not publish complete information about the names and positions of principal investigators and adequate analyses of grant recipients are not available. (The foundations do publish the names of recipients, but they do not list their ranks or administrative positions;[16] in 2006 the RFBR announced its intention to give more complete data in the near future.) Thus it is clear that while the foundations by their nature minimize the influence of the chief administrators of science, at least compared to the previous Soviet system, they do not entirely avoid this factor. Throughout the scientific world the most effective way to resist this natural tendency for powerful administrators to have great influence is recognized to be the rotation of peer reviewers and program officers in foundations, and also committee-based decisions making. Unfortunately, rotation of such personnel is rare in Russia; formally the foundations do make group decisions, but anecdotal evidence indicates that in final decisions there is heavy lobbying from prominent scientists and administrators, frequently board members, and therefore the ultimate decision is often the result of a compromise among lobbyists.

It is not surprising that the most critical respondents are those who have not been able to receive grants or who receive them rarely. Among these scientists the opinion is rather widespread that the foundations cannot support innovative ideas because, these critics allege, decisions are made on the basis of reigning theories and conventional concepts; therefore, ideas that do not fit within this body of ruling opinion are excluded. In other words, the foundations support "normal science" (according to Thomas Kuhn, whose work is well known in the Russian scientific community) and not "competing paradigms."

Critics like to say that it is impossible to make a scientific breakthrough by means of grant financing, and many of them doubt that the young Einstein would ever have been supported in that way. Obviously, such criticism is often self-serving and based on an exaggerated degree of self-importance (how many Einsteins are out there?) but may at the same time have some legitimacy. Nobel laureates in the United States have been known to make similar remarks about their difficulties in obtaining financing from NSF.

What are the chief positive influences of the foundations on the scientific community and the professional development of scientists? What are the chief problems in the activities of the foundations? Do the foundations influence the process of the reform of Russian science, and in what way? It is impossible to answer these questions through the application of rigorous methods such as quantitative analyses of publications. We can obtain statistics on grant financing that show where scientific talent is concentrated, and on its disciplinary, regional, and organizational structure, but we do not have answers to questions about how the foundations are influencing the distribution of talent and affecting these structural characteristics. In order to obtain more empirical information that might help evaluate the role of foundations, one of the authors conducted a survey.[17]

According to respondents the problems in the activities of the foundations that lower their effectiveness and undermine their work in the opinions of the scientific community are the following:[18]

1. Inadequate information about the activities of foundations, the selection of projects, and the reason for rejection of applications, i.e., inadequate "transparency."
2. Poor relations between the foundations and the public.
3. The existence of programs for which only applicants of a certain age are eligible (usually "the young").
4. Complicated application forms; paperwork required.
5. Complicated bookkeeping required for grants.
6. The absence of anonymity during selection; the experts know the names of applicants and base their selection on these names.
7. Delays and incompleteness in financing projects.

From the above list it is clear that the most significant complaint is that the foundations do not communicate well. As a rule the foundations in their public statements merely announce programs and supply application forms. Researchers complain that they do not receive information about who selects projects and how they are chosen, nor do they receive information on how and why the priorities and activities of the foundations are changing. Furthermore, the researchers are not satisfied with the information available about final results of competitions. There are many complaints about the predominance of grant recipients from a few large cities like Moscow and St. Petersburg.

Many of the complaints of Russian researchers about their foundations are similar to the complaints of scientists elsewhere, including the United States. There, too, researchers from the more remote locations often complain about the predominance of researchers from a few prestigious institutions among grant recipients. And there too scientists often believe that inadequate information is released by foundations about the reasons for the rejection of applications. (Some U.S. foundations, particularly private ones, give out no information at all on the reasons proposals have been unsuccessful; others, such as the National Science Foundation, on request give out anonymous evaluations written by peer reviewers.) But in Russia such complaints are probably more frequent and strong for several understandable reasons: the need for foundation support is not just to carry out research, but for the maintenance of a minimal living standard for scientists and their families; the number and types of foundations in Russia are very few; and the suspicion of manipulation and inside influence is high as a result of past experience extending back to the Soviet period.

At the same time Russian scientists seem rather unappreciative (at least to foundation administrators) of problems over which the foundations have no control and for which they should not be blamed. The foundation administrators are totally dependent on the federal budget and on the timeliness and completeness with which they receive funds. Both the administrators and the grant recipients experience great frustration over the fact that funds earlier promised by the government are often not actually produced (although the administrators personally suffer less from these shortfalls than the grant recipients).

Some of the grievances of researchers would be reduced if grants from the government science foundations were merely one among several possible supplements to block funding. But because these grants today are so important, and because for the individual researcher there is usually only one, or possibly two, potential sources for such government grants, the demand for them is great. And when the application of a scientist fails, the sense of injustice is often understandably high.

Predictably the level of criticism of the foundations is highest among those scientists who have applied to them a number of times but have never received a grant. These excluded scientists often maintain that the people who receive support are not those with the greatest talent for research but those who have a special ability to write convincing applications. (In U.S. science, such people are sometimes called grant swingers.) These same critics often say that the grant system favors those who are by accident of place of work already favored, causes instability in science, wastes the time of people who have to write applications, and thereby reduces the time available for actual research. Although much of this criticism can be explained as the grumbling of second-rate scientists, there may be a residue of legitimate complaint. The complaints are not very different from those heard in other countries. The unusual strength of such criticism in Russia comes, at least in part, from the memory of Soviet times when scientists' incomes and research programs were not dependent on competitive grants.

To sum up, the basic idea of competitive grant funding is not disputed today by Russian researchers, quite a change in attitude from ten or fifteen years ago. Displeasure, however, is still expressed about the procedures being used. And yet despite this river of criticism, a large majority (75 percent of scientists surveyed) speak of the positive influence grants have had on their own success in research, as indicated in table 4.4.[19]

As the results summarized in table 4.4 indicate, the chief positive effects of receiving grants are seen by their recipients as a combination of material support, involvement in the world scientific community, and growing professionalism. Natural scientists particularly value the opportunity to renew their equipment and buy laboratory materials and reagents. Humanists mention the ability to purchase needed literature, subsidize the publication of books, and study and do research abroad. All these forms of support facilitate the formulation of new scientific tasks, the exchange of experience with foreign colleagues (especially difficult in Soviet times), working with colleagues with fewer bureaucratic barriers, giving papers at conferences, and publishing one's work. Young people prefer to work in those laboratories and groups where grants have been received, since this success is considered an indicator of high-level research and of the possibility of quickly "enrolling" in the international scientific community. For a young person to have such possibilities in Russia often decreases the temptation to emigrate abroad.

Yet another gain among grant recipients was in the sociopsychological situation in scientific teams. It may be difficult to prove that grants actually cause scientists to do better research (although the circumstantial evidence for this is strong), but there is no question that they raise scientists' professional pride and self-respect, and the recognition they receive from colleagues. Many scientists say that grants bring reputation, independence, and publicity. In fact, since grants given by Russian foundations are very modest in size, scientists more and more turn to them as much for prestige as for material support.

For those scientists who desire to advance administratively, to become directors of laboratories or institutes, grants are not seen, according to surveys, as crucially important. The opinion is widespread that factors other than success in obtaining grants are more important in administrative promotion, such as age, place of work, status, personal magnetism, political connections, and even psychological constitution. Once a person obtains an influential administrative position, however, his or her name is rather often attached to an application for a foundation grant. As noted earlier, leaders of laboratories and institutes figure prominently among grant recipients. Thus, a curious phenomenon seems to emerge: working scientists do not think that success in gaining grants is important in being administratively promoted, but administrative leaders seem

Table 4.4. Changes occurring in the
professional careers of scientists who received grants

Changes	Frequency of response
Development of international ties, intensification of relations with foreign colleagues	19.4%
Better possibilities for purchasing equipment and reagents and consequent improvement in experimentation	14.6%
Growth in professionalism	7.2%
More freedom in scientific work	7.2%
Better access to information	7.2%
Growing respect from colleagues	4.9%
Better evaluation of one's own work	4.9%
Better possibilities for attracting more students and graduate students to projects	2.4%

Source: I. Dezhina, *Position of women researchers in Russian science and the role of foundations* (in Russian) (Moscow: TEIS, 2003).

to have unusual success in getting grants. Witnessing this phenomenon is one more reason that many Russian scientists combine respect for the grant system in principle with criticism of the way it works in Russia at the present time. (See table 4.5.)[20]

Respondents were also asked to indicate what directions foundations should move in the future (i.e., what are the priorities of funding?). Not many suggestions were given. Several were mentioned more than twice. The results were as follows, in decreasing frequency of response:

1. Increase money for purchase of equipment and modernization of telecommunication facilities, and also for Centers for the Shared Use of Equipment ("Centers for the Collective Use of Equipment," CCU).
2. Expand programs for travel expenses to conferences, including for undergraduate and graduate students.
3. Sponsor programs stimulating links between science and innovative industry.
4. Support libraries, both traditional and electronic ones; purchase periodicals and books.
5. Increase the average size of grants.
6. Finance conferences and seminars.
7. Give grants to young scientists and graduate students, up to age thirty.
8. Sponsor programs stimulating links between science and teaching.
9. Increase time periods of grants to five years.
10. Sponsor interdisciplinary research programs.

Table 4.5. Influence of the grant system and
foundations on administrative promotion and professional growth

Influence of foundations	Administrative promotion	Professional growth
Significant	27.1	78.3
None	58.9	13.3
Difficult to say	14.0	8.4
Total	100.0	100.0

Source: I. Dezhina, *Position of women researchers in Russian science and the role of foundations* (in Russian) (Moscow: TEIS, 2003).

As the list above indicates, respondents placed purchase of equipment and access to telecommunications at the top of their priorities. This desire, together with several others listed above, could be lumped together as a call for "better access to equipment and information" (including travel, links to industry, libraries). Interestingly, respondents ranked such needs higher than "increasing the average size of grants," indicating that the scientists valued the needs of research above their personal ones, even at a time when the average wages for scientists were very low. These priorities demonstrate the deep commitment of many Russian scientists to professionalism. The survey further shows that the future development of science in Russia is strongly linked by scientists to the foundations, probably more than to traditional government departments or ministries. Thus, the foundations bear a heavy burden, one they probably cannot continue to shoulder unless they are given more money.

The foundations—the RFBR and the RFH—understand their significance and are trying to fill critical gaps in funding. But, unfortunately, because the foundations have only modest resources to cope with such broad and diverse demands, the impression spreads among Russian scientists that every initiative the foundations undertake is incomplete and of low effectiveness. Thus, the effort to shift Russian science from dependence on the old centralized block-funded Soviet system to a new competitive one—based on the principle of individual initiative appropriate to a free-market economy—faces a severe test. Disillusionment could lead to greater nostalgia for the old system.

The development of government science foundations in Russia coincided with a crisis in science and was in part inspired by this crisis. It is only natural that the first response of the new foundations was to try to ensure the survival of Russian science; only later did they examine the secondary questions of development and reform. Now, however, the foundations are intimately involved in these secondary questions: by broadening international cooperation, improving research conditions, increasing freedom in research, elevating the professional standing of grant recipients, and stimulating a sense of responsibility by scientists for the advancement of their fields as well as their own careers, the foundations are among the most influential reforming forces in Russian science today. In that respect, their future is of fundamental importance.

Status of Foundations and Regulation of Their Activities

In Western countries there is no common approach to the problem of regulation of the activities of foundations. The legal bases of foundations differ from one country to another. In the United States, Derek Bok, former president of Harvard University, wrote in 1996 that of all the institutions in America "foundations are surely among the least accountable." And he further defended that status, adding that it would not "be wise to make them accountable in the conventional sense, for their business is to provide society's risk capital by supporting what other institutions have overlooked or find too chancy or uncomfortable to finance. Answering to outside groups might only make them more timid and reluctant to back controversial people or bold undertakings."[21]

Bok was speaking here primarily of private foundations, which are especially prominent in the United States. But even federal U.S. foundations, such as the National Science Foundation and the National Endowment for the Humanities, are given greater leeway than most federal institutions to make wide use of peer-reviewed grant awards. By and large they have preserved this status despite occasional clashes with the Congress, their funder.[22]

In several countries (Germany, Switzerland, and the Netherlands) science foundations are formed in accordance with civil codes, while in other countries special laws have been adopted for this purpose. Most national science foundations are government agencies and distribute funds from the state budget appropriated for research and development.

In Russia science foundations are in an ambiguous situation. The foundations began their work in the early 1990s when the basic laws and regulations for a capitalist economy did not yet exist in Russia. It is not surprising therefore that contradictions arose after a whole series of laws and codes governing economic entities was adopted. Some of these laws were devised specifically to regulate foundations. However, even now there is no legal clarity in Russia about the very concept and status of "government foundations."

The first of two foundations, the RFBR, was created in 1992 by the president of the Russian Federation, Boris Yeltsin, in response to the crisis of Russian science that the very title of his decree reflects: "Concerning Urgent Measures for Preserving the S&T Potential of the Russian Federation."[23] At this time Russia had no Civil Code, no Budget Code, and no laws regulating "non-commercial organizations." When these codes and laws were adopted later, incompatibilities with the previous presidential decree soon emerged.

In 1999 the Ministry of Finance initiated a reexamination of the charters of the science foundations (RFBR and RFH). It was proposed that the foundations choose one of two possible organizational forms, either that of a "government institution" or a "foundation." The problem was that the RFBR and the RFH possessed characteristics of both forms, and therefore the choice of one seemed inevitably to lead to the loss of certain desired characteristics of the other. Governance is one example of a major sticking point. On the one hand, these organizations have boards of trustees and operate on the principle of self-management, characteristics that make them seem more like "foundations" than "government institutions." On the other hand, the foundations are financed from the federal budget and their heads are appointed by the government, characteristics that make them seem to be government organizations or institutions. The two foundations were told that if they accept status as government institutions, financing from the federal budget is guaranteed, but then they cannot be self-governing organizations; instead, they must introduce central management in which the government is the final authority. Understandably, foundations have seen such a system of governance as a threat.

The foundations were also told that if they choose the legal framework of a foundation, then government financing is no longer guaranteed. On the other hand, according to the newly adopted Civil Code, foundations can possess property of their own and can carry out entrepreneurial activity, while government institutions cannot.

In 2001 the foundations adopted new charters according to which they became governmental non-commercial organizations; thus, they are now federal institutions under the supervision of the government that finance research and other activities of scientists. In the new charters the term "self-governing organization" disappeared, as did such concepts as "grants" and "individual projects." The concept of "research team" was replaced by "organization."[24] These changes meant a broadening of the administrative rights of the director's office in the choice of groups to perform projects. They also called into question whether the foundations' original emphasis on individuals, not organizations, would be retained.

In accordance with the Civil Code, the RFBR is an "institution created and financed by its owner (the government) for carrying out functions of a non-commercial character." Unfortunately, this definition does not provide answers to basic questions about whether and how the RFBR differs from most other government institutions: To what degree is it autonomous? Can it finance projects on the basis of independent (non-government) peer review?

Furthermore, the economic position of the RFBR and the RFH was greatly complicated by the adoption on December 8, 2003, of a new law according to which "federal ministries and agencies must be the chief managers of money from the federal budget."[25] It was not clear whether the foundations will in the future be considered to be such agencies, or whether they will be forced to get their money from a ministry, perhaps by means of some sort of "pass-through" mechanism.

An additional problem is that in Russian law the concept of a grant is still not clarified. There has been much discussion over the distinctions among "contracts," "grants," "cooperative agreements," and "donations." The closest term to the concept of a grant that is defined in the Civil Code of the Russian Federation is a "donation," recognized as "the gift of a thing or right to citizens, individual organizations, and the government." Whether this concept is workable for "research grants"—a term in common usage now in Russia, even if not legally clarified—is still uncertain.

At the present a draft law has been approved by the State Duma and the Federal Council exempting grants from Russian science foundations from federal income tax. The same tax regulations now apply to both foreign and Russian foundation grants.[26] This new law is a recognition by the government of the significance of foundations to scientists.

From the Russian standpoint, science foundations have both strengths and weaknesses. Their strengths include: (1) identifying and supporting the strongest projects, both from a scientific point of view and from the point of view of social utility; (2) overcoming departmental, regional, and disciplinary barriers; (3) integrating science with a market economy; and (4) providing recognition to outstanding scientists and research teams.

Their weaknesses seem to be: (1) foundations are not suited for, nor should they aim for, the support of all the needs of science and scientists, such as libraries, buildings, installations, and support personnel; (2) foundations are not appropriate or adequate for solving personnel problems; and (3) foundations often reveal weaknesses in scientific establishments (unproductive researchers, bureaucratic barriers), but they cannot automatically solve these problems.

The above strengths and weaknesses of foundations may be generic ones, but foundations in Russia also face some crucial specific problems. Perhaps most important is the question of the degree of independence possessed by the foundations. As organizations created by and funded entirely by the government, it is not surprising that they have now, after much discussion, been defined as government institutions (the NSF and the NEH in the United States, which served as models for the Russian foundations, are unquestionably government institutions). But just how much independence will the RFBR and the RFH have in the future? The effort by some Russian politicians to bring the foundations under the control of a ministry has disturbing implications, especially in view of the continuing hostility of some administrators to autonomous foundations (this ministry would probably be the one in charge of science, which has held various different names in recent years; at present it is the Ministry of Education and Science).

Another great weakness of the government science foundations has been that their budgets are small. The damage of these small budgets goes beyond what it does to scientific research in Russia. As the surveys we have cited indicate, Russian scientists increasingly link these foundations to their main hopes for an improvement in the situation of Russian science. The inability of the foundations to deal with a task of such large dimensions leads to growing dissatisfaction in the scientific community. That dissatisfaction could conceivably lead in one of two quite different directions: a backlash, a turn to hopes for old centralized block funding; or radical demands for a new type of scientific establishment, perhaps even more similar to Western models.

Gradualist reformers such as Boris Saltykov,[27] who want an improvement in the current situation with a continuing important role for the science foundations, have called for a number of changes. They include: (1) clarification of the legal status of the RFBR and the RFH so that they continue their work with a high degree of independence; (2) a large increase in the funds allocated to these foundations, perhaps by 2.5 to 3 times, supplied in a regular, stable way; (3) clarification, legally and financially, of the concept of a grant; (4) greater consultation with the scientific community about changing needs: equipment, travel grants, support of youth, centers for the shared use of equipment; and (5) promoting greater cooperation between federal and regional authorities in the support of research, especially in innovation technology.

Even though the history of government science foundations in Russia is short, already one can draw a number of conclusions. First of all, it is clear that the basic conception of these foundations differs in important ways from traditional forms of the government financing of science in Russia. Foundations are oriented toward the creator of science (scientists, research team), not the institution in which they work. Foundations prefer to finance concrete projects and not general themes. They rely on selection by independent peer review, not by administrative choice. They call for transparency in accounting and free access to research results. They assume that scientists should have equal opportunities to seek government research funds without interference from institutional authorities.

The rise of foundations of this type in Russia since the fall of the Soviet Union points in an unusually clear way to the relationship between the political/economic structure of a country and the predominant way in which that country promotes science. The one-party authoritarian political system of the Soviet Union, with its centrally planned economy, promoted science from above. (The fact that this science was sometimes rather good shows that while central direction is not the best system for science, it can survive central direction better than the economy as a whole, and explains in part the institutional conservatism of many Russian scientists.) After the collapse of the Soviet system and the adoption of principles of political democracy and economic capitalism, many scientists and administrators did not at first realize that the implications of these political and economic changes were as broad and deep for science as they turned out to be. What the scientists wanted most of all was secure financing. When the government turned to the foundations as a means of rationing scarce resources among scientists with needs greater than it could satisfy, the scientists were at first, quite understandably, suspicious and resentful.

With time, however, most of the scientists learned to like the new principles. Always (and accurately) worried that not enough money was available, they came to see that competition for money based on peer-reviewed applications was likely to result in the support of better quality research. That meant, of course, that some scientists were unsuccessful, and slid back into poverty or left for other lines of work. Adjusting to such Darwinian principles was not only painful but even contradicted much of the spirit of "collectivism" still strong in Russian life; however, as the survey results demonstrate, that adjustment has been occurring.

The questions, then, that have emerged among science administrators and scientists as the most important today are somewhat different from those asked only a few years ago. The most pressing questions at that time (early nineties) were "Can Russian science survive?" and "How should scarce resources be distributed during a crisis?" The most pressing current questions are "What is the correct balance between foundation funding and traditional block funding?" and "How much independence from government direction should the foundations have?" (Another important science policy question, discussed elsewhere in this book, is "What should the relationship be between research and teaching, between academies and universities?) The shift in the types of question being asked reveals important progress: Russian science *has* survived. Now the worries are about how it should be administered, funded, and best promoted under current political and economic conditions. These questions will probably remain controversial in Russia for some time. Indeed, they are controversial in many other countries today.

5 Developing a Commercial
Culture for Russian Science

Immediately after the collapse of the Soviet Union the first priority of scientists and science administrators in Russia was to preserve science itself, to save it from a similar disappearance. That task has been long and difficult, as shown in other chapters in this book, but some success has been achieved. Despite losses of many talented scientists to brain drain and dramatic cuts in research budgets, science in Russia continues to have a life. In recent years, however, another task has risen in importance, one based not on preserving the strengths of the past, but on creating something entirely new: a commercial culture for science that would help Russia to compete economically in high technology with other industrialized nations. This task has proved to be the most difficult of all those considered in this book, and whether Russia will solve it or not is still a very open question.

Soviet science was powerful in many areas of fundamental research, but a Western-style commercial culture was totally lacking. The Soviet Union had no private property, no legal system defending patent rights and intellectual property, no financial system or venture investing funds for promoting innovation activity, and no organizations or institutions facilitating the transformation of fundamental knowledge into commercially viable innovations and products. Even more importantly, it had no tradition of demand from business for innovation, no tradition of individual entrepreneurs in high technology, and very little knowledge of marketing skills in technology. The Soviet system was based on political command from above rather than economic pull from below.

In this chapter we will describe recent efforts in Russia to create an S&T Infrastructure, what the Russians now often call a National Innovation System (NIS). They define this NIS as a network of interconnected subsystems designed to provide money and services necessary for innovation activity. As will be clear, many difficulties have arisen in trying to do this. The hope of Russian political leaders has been that with a little help from the government this commercial culture can later take off on its own on the basis of economic pull from below, in both domestic and foreign markets.

The necessity for creating such a commercial culture for science in Russia was dictated not merely by the characteristics of the preceding Soviet economy and its lack of an S&T Infrastructure, but also by the nature of the emergent globalized economy. The Soviet Union could survive economically—at least for several decades—as a mostly self-contained economy. The new Russia might similarly survive for a few decades based on exports of natural resources such as oil and gas—the basis of its current new economic health—but it cannot prosper in the long run unless it joins in world trade in technology. The emergence of a global economy, in which not only established industrialized states such as those in Europe, Japan, and North America are major participants but also emerging nations such as China, India, Singapore, Taiwan, Malaysia, Brazil, and the Philippines, means that even countries with strong science traditions like Russia cannot compete unless they possess a commercial culture promoting technological innovation. Good science by itself is simply not enough in the current world economic scene.

Statistics reveal just how far Russia is from this goal of becoming technologically competitive: Russia's share of the international market in high technology at the present time is estimated as between 0.35 and 1 percent.[1] This share is lower not only than that of other developed countries in the world but even than that of the developing countries of Asia. The total trade in technologies (including patents and licenses) from Russia in 2001 was $636.9 million, and in 2002 it was $784 million.[2] The improvement in one year is encouraging, but Russia has a long way to go. The total trade in technologies from the United States in 2002 was $49.7 billion, and even little Switzerland in that year exported technologies worth $3.5 billion, almost four and one-half times more than Russia.

If one attempts to measure innovating activity by looking at the number of patent applications submitted per 10,000 of population, both in the domestic and foreign markets, the situation appears slightly better. Russia ranks at a middle level (2.62), ahead of countries such as the Czech Republic, Poland, and Hungary (0.6–0.7) but behind leading industrialized states with corresponding figures of 4.5–5.5.[3] The implication here seems to be that Russia does somewhat better at innovating than it does at marketing its innovations and making them competitive in world trade.

The Russian government has attempted to stimulate innovation in three ways: by forming or facilitating new financial mechanisms, primarily "funds" to help finance innovating businesses; by creating or facilitating new organizational structures to support innovation (Technology Parks, Innovation Technology Centers, Special Economic zones, Technology Transfer Offices); and, last, by creating new legal mechanisms to protect and promote innovation (laws on intellectual property, patent laws, and the like). In this chapter we will consider these three systems—funds, organizations, laws.

Before examining the details of these systems, however, we anticipate some of our conclusions by noting that so far their success in stimulating innovation has not been great. The number of successful innovative businesses in Russia is not appreciably growing, and only about 10 percent of the special organizational structures formed for the purpose of stimulating such businesses are in fact doing so.

In the Russian approach certain characteristics of early Soviet attitudes are often still visible: more emphasis on governmental push from above than economic pull from below, more weight given to quantitative results than qualitative ones, and relatively little attention to profitability as a criterion of success.

New Financial Funds

The first financial organizations supporting S&T in Russia arose shortly after the collapse of the Soviet Union. Their original purpose, however, was not innovation development *per se*, but instead the support of the existing vast scientific establishment, especially that part connected with applied research and development. Later several of them modified their activity to promote innovation.

In the Russian language the word "*fond*" can be translated into English both as "foundation" and as "fund." In English the words have different meanings. "Foundation" usually means an organization pursuing philanthropic purposes and giving grants, while "fund" is simply a pool of money that can be used for different purposes: a family fund, an investment fund, etc. Sometimes it is difficult to know which way the word is intended in Russian. Some of the organizations we will call funds (because that seems to fit better with common English usage) are called foundations by Russians and in other English sources.

The Russian Fund for Technological Development (RFTD)

The first of these organizations was the Russian Fund for Technological Development (RFTD), created in 1992 by the Ministry of Science and Technology Policy of the Russian Federation.[4] Some of the difficulties the RFTD subsequently encountered are connected with the date of its founding, very early in the post-Soviet period. In 1992 there was still almost a complete legal vacuum; laws allowing private business and non-governmental organizations were still not enacted. Therefore, legal clarity has not been a characteristic of the RFTD's operations.

Although the RFTD was initiated by the government, its budget comes not from federal funds but from voluntary allocations from industrial enterprises.[5] The size of the RFTD budget varies within the limits of 1.5 to 4 percent of the federal expenses for civilian science. The RFTD supports projects at the R&D stage, and most of its recipients are research organizations and small enterprises.

The basic tasks of the RFTD are to promote those "critical technologies" stipulated by federal priorities, to support R&D aimed at solving the most important social tasks of the country, and to help finance innovation projects and S&T infrastructure. The RFTD finances R&D in the form of no-interest loans for a defined period (usually no longer than three years). The final result of the project, after its successful completion, must be experimental or pilot-plant production.[6]

The types of industrial projects financed by the RFTD are determined to a significant degree by the origin of the money the Fund receives. Industries making allocations to the RFTD naturally promote topics of their own interest. In the first ten years of its existence the RFTD financed projects primarily in chemistry and new materials, machine and instrument construction, and information technology and electronics. Seventy percent of the money disbursed was for these areas, but in addition small amounts were allotted to almost all areas of applied science, including medicine, the power industry, ecology, aeronautics, biology, metallurgy, and agriculture.

Approximately 50 percent of all recipients of RFTD support have been joint-stock companies, usually small ones. About 20 percent have been Federal Research Centers (these Centers are discussed elsewhere; they are former industrial [*otraslevye*] R&D institutes that have been rated most highly by the government). Another 20 percent have been institutes in the Russian Academy of Sciences and other academies (e.g., medicine, agriculture). Universities make up only 5 percent of the recipients.

As mentioned earlier, the legal status of the RFTD is not clear. Recently a number of laws have been passed which, if not actually banning the work of the RFTD by the old rules, at least make it extremely difficult. The status of the RFTD as a government institution with a non-government budget puts it outside the framework of standard legal definitions. Up to the present time the Fund has not succeeded in finding its exact position in the law and in the innovation cycle.

Despite these difficulties, the RFTD seems to be successful in some of its work. One of its major strengths is its ability to support the best interdisciplinary research involving several different industries. Such projects are not often supported by individual enterprises, and thus the RFTD has an important function.

As of December 1, 2003, the overdue debt owed to the RFTD consisted of 7.8 percent of total loans given, a tolerable level. According to the RFTD each ruble invested by it yielded a profit of three rubles. Of the 840 S&T projects supported by the RFTD in the period 1994–2003, 39 percent reported an increase in market share after the completion of their projects. The successful applicants came from a variety of industries, including instrument construction, power production, electronics, biotechnology, chemistry, medicine, and pharmacology.

However, the RFTD continued to be plagued by legal problems. In 2004 these difficulties resulted in a temporary freezing of its activities. Then in the spring of 2006 it won permission to recommence its work.

The RFTD deserves additional study. While it has supported only a relatively small number of projects (less than a hundred a year) it seems to be a modest success in some of its activities. But its future is still not certain, and it continues to have periodic difficulties of an administrative and legal nature.

The Fund for Assistance to Small Innovative Enterprises ("the Fund for Assistance")

A second financial organization for promoting S&T is the Fund for Assistance to Small Innovative Enterprises, created in 1994 by the Russian government.[7] It is often called, for brevity, the Fund for Assistance. The budget of this fund comes from the government and is 1.5 percent of the total federal expenditures for civilian science. The Fund has a special line in the federal budget and is independent of any ministry, but the money is actually transferred through the Ministry of Education and Science. It supports small innovative enterprises that are at the stage of commercial production and, since 2003, new innovation companies (start-ups). It differs from the RFTD in supporting technology at a later stage of development; the RFTD supports R&D up to the stage of pilot or experimental production. The Fund for Assistance begins its support when the company is actually producing commercial products. Nonetheless, there is some overlap in the two functions.

Two foreign programs were important models for the Fund for Assistance: a U.S. program, promoted by various departments and agencies of the U.S. government, called Small Business Innovation Research (SBIR),[8] and the French state agency for innovation (*Agence Nationale de Valorisation de la Recherche*, or ANVAR).[9] However, the Russians did not simply copy these foreign models, but drew upon different features of each. From SBIR the Russians took the idea of government money in support of small innovative businesses, and from ANVAR came the idea of organizing a set of regional representatives. At the present time representatives of the Fund for Assistance have been created in twenty-five regions of Russia.

Both SBIR and ANVAR function on the basis of support through open competitions, peer review, and decision making by an independent jury (academic council). These same principles underlie the work of both the Fund for Assistance and the RFTD.

The small innovative companies that the Fund for Assistance supports have the rights to the intellectual property (IP) they create. The financial share of the Fund for Assistance in the winning projects can not exceed 50 percent.

The Fund for Assistance has supported a great variety of businesses, ranging from pharmacology to machine construction to agriculture to computer technology. Most of these businesses were already established and operating. The money was given out as loans at preferential rates. The strategy was passive in the sense that the Fund followed developing enterprises and did not directly stimulate completely new products. Working in this relatively low-risk way the Fund was often successful, with an average level of return on credits of 66 percent.[10]

Recently the Fund for Assistance has begun to support companies at the seed and start-up stages, obviously a riskier strategy. The stimulus for this change derived not only from a stabilization of the general economic situation, but also from the fact that the financial resources of the Fund were becoming larger. Federal appropriations to the Fund increased from 0.5 to 1.5 percent of the expenditures on civilian science, and at the same time the overall budget for civilian science grew in absolute terms. Also, the return to the Fund from earlier loans provided additional money.

The next step was support for innovation projects in the very beginning stages on the basis of a new program, called START. Approximately half of the Fund's budget was devoted to the START program ($10 million when the Fund was started in 2003; in 2006 approximately $15 million).

The START program reminds one by its structure of the U.S. SBIR program, but in Russia's case the step between R&D and a prototype must be taken very rapidly, in a year. The risks involved in such a short transition are high.

In the cases of both the RFTD and the Fund, an important difference between the Russian efforts to promote innovation and the U.S. procedure is a feature that does not show up clearly in the official documents. In the case of SBIR (and also the French ANVAR), governments predicate their efforts to help innovative enterprises on the assumption that private investment capital will move in once initial success is demonstrated. The U.S. and French efforts can best be described, therefore, as "pump-priming" until more powerful economic forces take over. In the Russian case, despite the modest successes gained by the RFTD and the Fund for Assistance, this later stage is often missing or only weakly expressed.

Furthermore, the basic criteria for defining the effectiveness of the work of the Fund for Assistance do not place much emphasis on profitability, but instead employ quantitative measures that seem somewhat similar to old Soviet standards: rate of growth of production of small enterprises (must be more than 15 percent a year), output per person per year (must be at least $20,000), and the amount of newly created IP that has been transformed into commercial products. By these criteria, about 50 percent of the small enterprises supported by the Fund for Assistance have been successful. Unfortunately, these criteria, interesting and helpful though they may be, do not give a very clear picture of the profit-making abilities of the enterprises and therefore are rather silent on the question of sustainability.

Just as is the case with the RFTD, the legal situation of the Fund for Assistance is complicated. It is a government institution and transfers government funds to non-government enterprises, to which it also assigns rights to new IP. However, there are contradictions between the Civil Code and the Budget Code of the Russian Federation on whether such activity is legal. According to the Civil Code foundations can make such grants, but according to the Budget Code government institutions cannot transfer IP rights to non-government organizations. This problem is unresolved. A new government resolution of November 2005 simplifies the situation somewhat, but does not completely resolve it.[11]

The Venture Innovation Fund, and the Development of Venture Industry

The Russian government formed the Venture Innovation Fund (VIF) in 2000.[12] The key initiator for this Fund was the former Russian Ministry of Industry, Science, and Technology. The VIF is a non-commercial organization which, with government assistance, forms industrial and regional venture funds on the basis of long-term shared financing. A model for the VIF has been a "fund of funds," the Yozma Fund in Israel, which, in the span of three years, created ten regional funds. The budget of VIF comes from an earmarked appropriation from the Ministry of Industry, Science, and Technology, an earmarked appropriation from the Fund for Assistance taken from returns on loans, and also—and here is a potentially promising feature—voluntary amounts given by local investors. However, very little money showed up from this latter source, pointing to a serious weakness in all Russian efforts to develop an innovation industry. Originally the budget of the VIF was R200 million; in 2005 it was equal to R300 million (very roughly, $10 million), which for venture investing is so small that it is more symbolic than real financial support.

Not surprisingly, the VIF has not been very effective so far. Only two regional funds have been created, and the money available to them is so meager that they are more demonstration models than economically significant organizations.

* * *

Looking back at the efforts made by the three innovation funds that we have considered—the Russian Fund for Technological Development (RFTD), the Fund for Assistance for Small Innovative Enterprises, and the Venture Innovation Fund (VIF)—we can identify several reasons why they have had only very limited success. First of all, it is clear that the Russian government does not wish to assume real risks and therefore assigns little money to foster innovation. Second, there are few strong stimuli for investing in high-technology projects, either for the government or private individuals, at a time when it is significantly safer to invest in resource-extracting industries, such as oil, gas, and minerals. These extractive industries are the genuine strengths of the current Russian economy. (In January 2005 Russia actually produced more crude oil per day than Saudi Arabia, and petroleum exports represented almost 55 percent of the country's output.)[13] Thus, the strength of extractive industries depresses the position of innovation industries, soaking up much of what capital is available. To overcome this obstacle would require a major change in economic incentives, which so far the Russian government shows little sign of implementing.

Other obstacles in Russia to venture capital investing are the lack of experienced managers of venture capital; an undeveloped securities market; the short time horizon of those investors who do exist; weak protection of intellectual property; the unwillingness of many innovators (scientists, engineers) to give up controlling interests in their innovations; and the red tape and legal obstacles involved in complying with government regulations.

As a result of all these obstacles, those few venture capitalists who are active choose their investments, as a rule, on the basis of family connections or friendships rather than through an analytical approach to all possibilities existing at any given time. Furthermore, the unreliability of available statistics often makes determining the profitability of past investments difficult.[14] Not surprisingly, few large investments are directed toward new companies, finding their way instead to already existing ones in second or third rounds.

Nonetheless, some improvements in the sphere of venture financing are occurring. In 2002 a government directive for the first time mentioned the concept of venture funds, namely the directive of the Federal Commission for Securities (FKTsB) titled "Regulation of the composition and structure of assets of joint-stock investment funds."[15] This was a first step in showing the government's support for such funds and in listing the regulations governing them. A second improvement has been the clarifying of regulations on intellectual property, yet to be discussed in this chapter. A third positive tendency is a perceptible but still small growth in interest in venture investing among Russian private businesspeople and investors.

In retrospect it appears that the financing by the government of venture initiatives of the types described in our three funds was somewhat premature at a time when appropriate economic, legal, and administrative conditions were still absent. At the present time several government agencies are showing new interest in high technology venture funds.[16] It is still too early to judge the promise and sustainability of these efforts.

New Organizational Structures

Technology Parks

Technology Parks have been created to encourage cooperation between research institutions and industry with the goal of effective use of new technologies. This particular type of organization arose at the very end of the Soviet period, at a time when private enterprises were first encouraged. The Technology Park concept is in part borrowed from foreign experience. Most of them have been created at the initiative of universities or research institutes.

The first Technology Park was created in 1990 in Tomsk. One year later, in 1991, there were already 8 Technology Parks; in 1992 there were 24 and in 1993 there were 43. Some of these were more labels than realities, however, and the number of effectively operating Technology Parks remained small.

The main goal of Technology Parks is assisting the formation and growth of new science-intensive firms, using the results of research performed in institutes and universities. At the same time, the Technology Parks are intended to improve the local economic situation and stimulate new funds for their parent organizations.

At present there are about eighty Technology Parks in Russia, although an evaluation of their effectiveness in 2000 indicated that only about ten or twelve of these met international standards.[17] The evaluation was based on such criteria as strength of linkages with the parent university, level of involvement of students, number of technologies created and transferred to industrial enterprises, and degree of interest in the Technology Park from regional authorities and industries.[18] (Notice that these criteria contain no reference to profitability of the new innovations developed in the Technology Parks.)

The highest aggregate grades were given to ten Technology Parks, with the top grade given to the International Science-Technology Park created by the

Moscow State Engineering-Physics Institute. In addition to this institution, three others of the highly ranked Technology Parks were located in the larger Moscow region (Moscow State Power Institute, Moscow Institute of Electronic Technology, and Obninsk Institute of Atomic Energy). The St. Petersburg region, surprisingly, had only one (St. Petersburg State Electro-Technological University). The other five were located (in descending order of grades) in Saratov, Ufa, Nizhnyi Novgorod, Ulianovsk, and Tomsk.

Most Technology Parks in Russia have remained structural subdivisions of universities. The length of time that a firm may remain within a Technology Park is not limited, and currently is on average about ten years, compared with two to three years in Technology Parks outside Russia. Since many Russian Technology Parks are not much older than ten years themselves, whether firms will ever leave in large numbers is still an open question.

Part of the reason that many Russian Technology Parks seem to be functioning in a different way from those outside that country (on which they were, at least in part, modeled) is that Technology Parks in Russia bring different benefits to the firms within them: comfortable facilities (communications, consulting services) and protection from criminality. Most Russian Technology Parks are situated on guarded territory and are relatively well protected from the criminality and corruption that were rampant in the nineties and still exist. Important questions in the near future will be whether business crime diminishes, and whether the function of Technology Parks can become more similar to those abroad.

The small number of actually operating Technology Parks in Russia can be explained by the fact that they were formed not so much in response to perceived market needs as to receive government subsidies. The government did not base its selection on calculations of the commercial viability of the projects. And this insensitive government policy continued, as shown by the fact that the evaluation in 2000 mentioned above did not have any consequences—not on tax advantages nor on subsidies. Government money continued to be distributed uniformly to all operating Technology Parks, which were created with the assistance of such money.

The Technology Parks that have been successful are often those whose managers have been professionally trained, often abroad, where they studied Western experience. In these successful Technology Parks undergraduate and graduate students are involved, small firms have been formed, and the linkage to the base university is rather close.

Successful Technology Parks end up not as recipients of funds from their parent universities, but as donors to them. For example, the Science Park at Moscow State University not only pays the university rent for its land and for utilities (at a rate 36 percent higher than the city rate), but also rents at its own expense four hectares of surrounding land. It also produces scientific instruments for the laboratories of the university and supports external R&D. But here again the practice is somewhat different from the original intention of Technology Parks: this particular park contains established companies, often in information technology, instead of start-up firms. These companies do not seem to be interested in leaving the Technology Park and operating in the normal economy.

An example of a successful Technology Park which seems to be coming closer to the original goals is that of the Moscow Institute of Electronic Technology (MIET) located outside Moscow city proper in Zelenograd, an area that has sometimes been called Russia's Silicon Valley. In the middle nineties successful companies in this Technology Park grew and made their exit to the "free economy." In 2002 a group of about sixty science-intensive industries (both Russian and foreign) in Zelenograd launched an organization called a Technological Village, a new phenomenon in Russia. Foreign industries were interested not only in high-tech research but also in promoting their instruments and equipment among Russians in a way that they, in return, recommended these products to others. This Village occupies an area of around twenty thousand square meters and concentrates on electronics and information technologies.

The Technology Parks have some problems, particularly those connected with property, or expansion of territory. Most of them exist on university or research institute property, and there seems to be no legal way in which they can purchase the property from the academic institutions with which they are affiliated. Their experience shows that it is usually easier to expand by new construction on empty land or by completion of unfinished structures than it is to expand by transfer of unutilized university property to the Technology Parks.

In order to attempt to overcome these problems the Ministry of Economic Development and Trade developed in 2005[19] a new concept of the Technology Park, including two major improvements: (1) a larger territory where all facilities and equipment are located, and (2), "incubators" inside Technology Parks that will help launch start-up companies.

Innovation Technology Centers

The first Innovation Technology Center (ITC) was created in March 18, 1996, at the Svetlana plant in St. Petersburg, one of the leading enterprises of instrument construction in the former USSR. The ITCs were created by the Russian government (through the united efforts of the RFTD, the Fund for Assistance, and the ministries of science and education, which were separate at the time), and they were allotted significant financial resources—in 1997, equivalent to about $50 million. Although somewhat similar to the Technology Parks described in the preceding section, ITCs were designed differently: they were supposed to be located primarily in industries (not at universities and research institutes, as is the case with Technology Parks), and they were intended to support the development of innovative enterprises that had already passed through the most difficult stages of survival.

However, since strict criteria for forming ITCs were not established, in practice about 45 percent of ITCs were created in universities, not rarely on the basis of already existing Technology Parks, so that these two forms of S&T infrastructure became interwoven and seemed to be duplicative. In several instances conglomerates were formed uniting a Technology Park and an ITC, as was the case at the Moscow Institute of Electronic Technology (MIET) mentioned above. The Science Park of Moscow State University, on the contrary, was transformed from one form to another and became an ITC: its older name, Science Park of MGU, was simply transferred to the ITC. One of the strong incentives for the creation of ITCs was the acquisition of governmental funds for their support.

Currently in Russia there are 52 ITCs that unite more than 1,000 small firms. While this number may indicate progress, the overall development is still small. Russian analysts have noted that in Germany there are more than 300 organizations analogous to the Russian ITCs (although it should be noted that the criteria for comparison are somewhat unclear; some organizations in Russia have the status of ITC only in a nominal sense). ITCs render a variety of services to the small enterprises located in them: rental accommodations and also technical, informational, and consultative services, as well as formal and informal help in the search by small enterprises for funds for further development.

The sources of financing of ITCs significantly differ, and vary from 100 percent government support to 100 percent non-government support. Rent payments are a main source of financing for the Science Park of Moscow State University (along with a small amount from consulting), and also for the ITC of the Center of Photochemistry of the Russian Academy of Sciences. The share of government support for ITCs has declined in recent years.

The number of enterprises located in ITCs and Technology Parks is growing slowly at present. Demand for high-tech products inside the country remains low, and consequently there is no potential for a large number of new small high-tech firms. Industrial enterprises often buy new technology abroad and in that way obtain not only new products but also secure post-purchase service and maintenance. The majority of domestic small firms can not afford to present analogous services.

Those small enterprises already located in ITCs and Technology Parks are usually satisfied with the relatively comfortable conditions they find there, and do not try to grow in size and leave. In an attempt to stimulate the exit of firms that have "settled in," a number of ITCs have raised rental rates, but the firms usually agree to pay more and remain in their earlier places. This tendency to defeat the original purpose of ITCs and Technology Parks is self-reinforcing because the parent organizations need rental income and are reluctant to push enterprises out in the absence of new applicants wanting to come in.

Nonetheless, the ITCs have had some successes. According to a survey conducted in 2001 at the request of the Fund for Assistance, the volume of goods and services produced by small enterprises located in ITCs is more than three times that of small enterprises not in ITCs, and the taxes paid by these firms in the course of three years reimbursed the original government investment.[20] These ITCs have been given a new impetus by the creation in 2005 of the new concept of Technology Park described on pp. 74–76. Some of these new Technology Parks may be connected with ITCs.

Special Economic Zones (SEZs)

Recently the government has promoted Special Economic Zones (SEZs), which provide on limited territories both direct and indirect stimuli for innovation. There are two types of SEZs, distinguished by focus on either production or high-tech development. The main stimulus would be favorable tax rates. The federal government plans to co-finance these zones at a 50:50 ratio together with local authorities. The federal government also has the right to define what types of activity will be permitted on these territories. Usually the zones are created on open spaces belonging to federal or local (municipal) authorities. The pattern for creation of SEZs seems to indicate that they will usually be located near industrially developed areas, and therefore they are a form of support for regions that are already strong and successful.[21]

After the adoption of the law creating Special Economic Zones, a federal agency was established to conduct competitions for their creation (so far, six high-tech development zones and two production SEZs have been created[22]). An evaluation of the work of the first SEZs is planned no earlier than 2008; after this evaluation a decision will be made about continuing the program with further competitions. Currently a discussion is ongoing in Russia over whether SEZs are appropriate. The SEZ approach has been widely applied in China and India, and the idea was adopted from there by Russian government administrators. Some Russian critics say that the creation of SEZs is an old-fashioned policy more fitting for developing countries than it is for Russia, which already has many advanced industrial areas. Such observations probably underestimate the current vitality of Chinese and Indian high technology industry.

Technology Transfer Offices

Despite the ambiguity of current Russian legislation on intellectual property, the government has begun to form mechanisms of commercialization. One of the first such mechanisms was Technology Transfer Offices (TTOs) in Russian research organizations. These are offices with the specific duty of fostering the commercialization of research by bringing research scientists together with business people.

In the United States, TTOs usually fulfill the following functions: drawing up patent applications for inventions; paying the costs, at least partially, of patent applications and annual fees for patents; issuing licenses for patented intellectual property; defending property rights from possible violations; collecting royalties from license holders; and distributing royalties in accordance with established agreements that usually include payments to the parent organization, to the inventor or inventors, and to the TTO itself (for administrative expenses).

Western models of such offices, especially in the United States, have been very influential in Russia. In the United States since the passage of the Bayh-Dole Act in 1980, research universities have been active in the creation of TTOs. University officers in the United States engaged in technology transfer have formed an Association of University Technology Managers (AUTM), which has worked with Russian officials. Here we have a curious example of double transfer to Russia: the "transfer of knowledge about the transfer of technology."

The Russian government has been particularly drawn to the idea of Technology Transfer Offices because it believes that such offices may become important suppliers of university budgets. Sometimes Russian administrators exaggerate the financial benefits of TTOs. In the United States, successful TTOs usually bring in from 0.5 percent to 2 percent of the annual budget of the university in question.[23] Nonetheless, great interest in TTOs has been displayed in Russia. In fact, in some cases enthusiasm for this idea has exceeded that of the Americans who taught them about it. There are critics of the effects of the Bayh-Dole Act in the United States, people who fear the effects of commercialization on the educational function of universities, but this criticism gets little attention in Russia.[24] In addition, there is inadequate appreciation in Russia of the fact that in countries such as the United States questions of intellectual property are still being hotly debated, especially in areas like digital technology.[25] In Russia there is a tendency to emulate "finished systems" in other countries, not understanding that these foreign systems are not finished at all, but are constantly evolving amid sharp controversies.

Even as the Russian government and university officials become more aware that TTOs are not likely to bring large royalties to universities, their enthusiasm for them remains high. Government administrators and economists hope that as a result of commercialization new small and middle-sized high-tech enterprises will be formed, creating well-remunerated positions for highly qualified workers and at the same time increasing tax income to governments as a result of additional economic activity.

TTOs in Russia have a pre-history. In Soviet times "patent departments" existed in R&D institutes and higher educational institutions. In the Soviet Union patents, properly understood, did not exist for Soviet citizens (although they did for foreigners). Soviet citizens were instead given "author's certificates," which entitled them for certain awards if the innovation was deemed valuable, but not royalties based on commercial use. The task of the old Soviet patent offices was the evaluation of innovations awards to inventors, and then, in some cases, making applications for patents for use by the institutions in foreign trade. Some of these functions are still retained by patent departments in Russian research organizations. Their old habits continue, and they usually do not aggressively seek patents or markets. Reforming these old Soviet-style patent departments has been slow. Though some of these older offices are now called Technology Transfer Offices, they are not usually very effective economically.

Two new types of TTOs have been introduced. The first type was begun in 2002 by the Russian Ministry of Industry, Science, and Technology.[26] The Ministry intends to create TTOs in "all leading government scientific organizations." Already by 2006 there were sixty-eight of this type of TTO. Since all universities in Russia active in research in science and engineering are government institutions, the call was addressed to universities and other research organizations. Unfortunately, the growth in the number of TTOs has not been accompanied by a commensurate increase in their financing by the government. In 2003 the average budgetary support for one TTO was R3 million, but in 2005 the average was R1.7 million.[27]

A second type of TTO is only in universities and is supported jointly by the Ministry of Education and Science and the U.S. foundation CRDF (Civilian Research and Development Foundation) under the Basic Research and Higher Education (BRHE) program. These TTOs serve first of all the interests of the university, and perform both commercial and educational tasks. They will be discussed more fully in chapter 7.

The presence of two models of TTOs is probably a positive fact and is in line with the majority of developed countries of the world, where there is no unified scheme of TTOs. The coexistence of various structures may maximize commercialization of technologies.

The TTOs have proceeded with commercialization in three directions: patenting (licensing), establishment of small (spin-off) companies, and R&D contracts with industrial enterprises. At the same time, they have become places for the training and retraining of innovation managers. More details are provided in the next section.

An Analysis of the First Results of the TTOs

Most of the new TTOs have been at work for only a short time—about three or four years—and therefore it is too early to expect significant results. Nonetheless it is interesting to look at the existing practices, which turn out to be heterogeneous, and also at the problems and obstacles that the TTOs have encountered.

The offices created with the support of the Ministry of Education and Science and the CRDF are developing rather successfully. Most of them were able to create the necessary pre-conditions: the presence of managers who were prepared and had some experience in commercialization, and also of patent attorneys and other lawyers. Despite this promising start, the university TTOs do not have a guaranteed permanent life because they have not yet gained significant income from commercial licensing. In fact, the share of their budgets coming from consulting and similar services does not exceed 5 percent. In the second year TTOs managed to establish stronger linkages with industrial enterprises by helping them with IP issues, providing consulting, and subcontracting R&D for industries' needs.

A problem, both in the TTOs created in the universities and in the independent ones created by the Ministry of Education and Science, is the qualification of the personnel in the practical work of commercialization. Such qualifications are improving but remain inadequate. Staff members of the TTOs are still inexperienced in working out patent protection of technologies and their subsequent successful commercialization. Learning these skills and completing innovation projects does not come quickly.

Not all the fault lies with the staffs of the TTOs. The scientists who are their clients often dislike engaging in this kind of work. Many of them seem to believe that working for economic success is not a proper role for a scientist. And those who are interested in such activity sometimes display an opposite extreme attitude: the belief that they can do everything on their own, and that the help of the TTO can only be as a source of additional financing. Although such attitudes exist in all countries, they are stronger in Russia than in those countries with long experience in, and appreciation of, the importance of commercialization and the important different roles played in it by scientists, advisors, lawyers, and investors.

A related problem is that scientists often prematurely propose R&D results for commercialization. As a rule they have created a model but not a finished product (including a prototype, full documentation, and so forth). Scientists often have an inadequate appreciation of economic feasibility.

Legal problems, especially in regulating contractual relationships between researchers and business people, complicate the work of TTOs. Inexperience is one source of these difficulties, and another is inadequate protection of intellectual property. (The TTO of Nizhnyi Novgorod State University had to do much basic spadework on IP, and worked up about 100 documents governing accounts, transfer of rights, inventory, etc.)

The TTOs at various universities and research institutions have poor communications with each other and do not share information adequately. The staff members of the TTOs are still trying to create professional networks.

It is still too early to evaluate the relative effectiveness of the two models of new TTOs, but some evidence seems to indicate that the model of the Ministry of Education and Science has been less successful because guaranteed support was given to these TTOs only for one year. (Expert evaluations indicate that only about 10 percent of non-CRDF-supported TTOs are developing successfully.) TTOs created under the BRHE program receive stable support at least for three years and thus they can plan their activity within a normal time frame, including time-consuming training and retraining fellowships for TTO staff members.

New Legal Frameworks

Unclear and insecure guarantees of intellectual property are a major obstacle to the creation of a thriving commercial culture for science and technology. Without such guarantees investors are usually unwilling to provide capital for technological development.

In Russia a related central issue has been: "Who has the rights to intellectual property created with federal funds—the government, or the actual creator of the IP (whether an individual or a research organization)?" This question is important in all countries, but it has deeper significance in Russia than it would in a country like the United States, since in Russia almost all research in science and technology has traditionally been funded by the government. But even in the United States opinion on this question has not been uniform; not until the passage of the Bayh-Dole Act in 1980 were the rights to IP created in universities by federal funds transferred to the universities themselves, which could further license or sell such rights to commercial enterprises (the government retained certain rights for its own non-commercial purposes). The goal of the Bayh-Dole Act was to promote the cooperation of universities and business in the commercialization of science and technology. The new post-Soviet Russian government has similar goals. Yet it has had considerable difficulty in devising legislation that would enable their attainment.

On the question of IP one should distinguish two different processes: the obtaining of rights to commercially significant R&D results financed by the government, and the actual disposal of those rights. If a government or an institution holds the rights to a given piece of IP but does not introduce it into the market, it brings no kind of income and eventually becomes obsolescent. Furthermore, since patenting abroad and supporting such a patent are significant expenses, non-utilized but patented IP is an economic loss rather than a gain. Therefore, the final goal for the commercialization of industrial innovations and technologies is the effective transfer and distribution of IP from R&D organizations to market production.

In the period from 1992 to 2002 a complex set of laws, edicts, and legal acts was adopted in Russia concerning rights to IP.[28] This legislation became stronger from a juridical standpoint and was an important step toward establishing traditional global practices. Russian legislation in the area of IP was borrowed largely from U.S. and European legislation, and it (incorrectly) assumed the existence of an operating market economy with clearly defined, stable property relationships. Nothing was said in the laws about the distribution of rights to IP created by government funds in case of privatization.

The Patent Law of 1992 was adopted when privatization of industry had not yet begun. Therefore a significant number of the research institutes, industrial enterprises, and innovation firms that received rights to IP on the basis of that law were, as earlier, property of the government. In that way, even though rights to IP may have been transferred to the enterprise or research institute, the government remained directly or indirectly its owner. After privatization it was very difficult to define the real owners of IP that had been created earlier.

On military and dual-use technology the position of the government was much clearer. According to the legislation, rights to this type of IP obtained at the expense of the government belonged entirely to the Russian Federation.[29] However, in 1998 and 1999 new legislation seemed to broaden the definition of what constitutes defense and national security in a troubling way. A special resolution of the government on September 2, 1999, stated that exclusive rights to any results of S&T activity paid for by the government were assigned to the Russian Federation unless the exclusive rights had earlier been elsewhere assigned. An important difference could be noted between the legal acts of 1992–1993 on IP and that of 1998–1999. In the earlier legislation, organizations had the title to IP and the rights of the government were undefined. In the legislation of 1998–1999 the assertion of government rights to IP was already the norm, with the exceptions being those cases when "rights had earlier been elsewhere assigned." What this meant in actual fact was still unclear.

Since the majority of objects of IP continued to be created with the financial help of the government, the administrators of most research organizations remained uncertain about their rights to it. A beginning to a clarification and normalization of the situation came at the end of 2001, when the government of the Russian Federation issued a new directive designating it as retaining rights to IP created at government expense when those rights "are directly connected with defense and the national security of the country, and also those for which industrial application is a task assumed by the government."[30] In all other instances the rights of the government to the results of R&D would be transferred either to the organization that developed the innovation, or to another economic agent.

The government of the Russian Federation is continuing at present to make efforts to develop legal policies on IP more in line with world practice. At the end of 2005 an important step was taken toward modernizing the legal regulation of intellectual property created by government funds. On November 17 of that year the Russian government adopted a resolution "Concerning the distribution of rights to the results of R&D." This resolution strengthened rights of research organizations to the results of R&D performed at government expense.

The new regulations say that in government contracts rights to the results of R&D can, in different circumstances, go either to the state, or to the organization where the IP was created, or to both. If the results are restricted in their use by the government (for example, certain types of biological cloning), or if the government is ready to be responsible for them, then the IP belongs to the state. If the results are defense-related or are related to government functions such as health protection, then both the state and the R&D organization may own the IP.

A two-year period of uncertainty about IP may be approaching its end. The recent legislation is still unclear on certain points, but it represents a big step toward a more liberal position on IP.

An Attempted Model for Transferring Intellectual Property to the Market

In 2002 the Russian Foundation for Basic Research (RFBR) and the Fund for Assistance attempted to create a model scheme for the transfer of intellectual property from the government to private industry. They based their scheme on similar mechanisms used in developed countries (Canada, the UK) in which government funds not only support fundamental research but also aid the commercialization of their results. The RFBR and the Fund for Assistance announced a joint program for the support of innovation projects on the basis of money coming from three different sources: the RFBR, the Fund for Assistance, and small private firms. The hope was to unite researchers who earlier had grants from the RFBR with small firms that were prepared to invest in products ready for market. Selection of projects was made by a panel of experts including scientists with experience in applied R&D. Each project, depending on the scale, received financing from the two agencies equal to R1.5 to R3 million.

The issue on which the attempted model ran into trouble was, predictably, intellectual property. The position of the two government agencies was that they must transfer rights to IP first to the inventors, who would in turn, on the basis of written agreements, transfer the rights to the small firms. A legal problem arose: the RFBR always requires that the results of its work (support of fundamental science) be published in the open press. Not all small firms, wishing to protect commercial innovation, wanted that to happen. Nonetheless, this attempt was made, and approximately half of supported projects turned out to be successful, showing that the potential for fruitful innovation was present. ("Success" was defined as sale of innovative products on the open market.)

However, the government auditing agency (the Accounting Chamber) concluded that such a program was not legal, since, according to the Budget Code, the RFBR is only able to spend money on fundamental research and cannot finance R&D at small enterprises. Therefore, use of the model was not attempted in the following year. Here again we see that the current legal system in Russia hinders the development of links between participants in the innovation process. Assignment of IP rights to organizations is still difficult.

The regulation of rights to IP in cases of partial financing of R&D by the government is a special problem. The most common proposed solution in Russia is joint ownership of such rights, on the basis of an agreement among the partners, taking account of their shares of the financing of the research. However, some observers have noted that it may be more advantageous to the government simply to transfer its rights to the performer of the R&D or to industry, since such a transfer may result in payment of more taxes.

At the end of 2006 the Russian government sponsored two new efforts in venture financing, which rather dramatically changed the existing situation. For the first time, large amounts of government money were dedicated to promoting high technology and, at the same time, a plan was advanced to attract private investors to this activity.

The two new organizations are the Russian Venture Company (RVC) and the Russian Investors Fund for Information-Communication Technologies. Both are based on foreign experience. The RVC is an open share society initiated by the government to manage a Fund of Funds, which was assigned R15 billion (about $600 million U.S.). The Fund of Funds will select high-technology companies for support. The Russian government also established conditions for private investors to join in this effort. Only between 25 and 49 percent of the total investment made in such high-technology companies will come from the government; the rest must come from private investors.

The second of the two new organizations, the Russian Investors Fund for Information-Communication Technologies, is more sharply focused than the Fund of Funds, as its title indicates. This fund will receive government money for a term of only three years; after, that private investors will take over. At the end of 2006 the government assigned R1.45 billion (about $60 million U.S.) to this fund, and another R1.5 billion will come from private investors. Thus, from the beginning, the amount of government funds given to information technologies under this program will not exceed 49 percent.

These new efforts to stimulate high technology enterprises with venture capital are significant in two ways: first of all, the amounts of money involved are large, and second, the Russian government has at last given the problem of venture capital much more emphasis and attention.

How successful has Russia been in promoting technological innovations, using all three of the mechanisms discussed in this chapter (new financial funds, new organizational structures, new laws)? Despite individual successes in such programs as ITCs and Technology Parks described above, the number of small innovative businesses in Russia has been steadily decreasing for the last eight or nine years. In 1996, according to published data, there were 46,700 such enterprises, and in 2004 only 20,700.[31] (These statistics are not entirely reliable, and the total number of such enterprises is probably larger; nonetheless the decline is a reality.)

The explanation of this perhaps surprising decline can be found in the recent economic history of Russia: soon after the collapse of the Soviet Union quite a few aspiring entrepreneurs broke off from the larger state organizations where they previously worked and tried to create their own enterprises; after exhausting the technical expertise they brought to the effort without achieving financial success, many returned to the larger organizations they earlier left, or dropped out completely. Some also left private business because of the stricter regulating legislation after the enactment of the Civil and Budget codes. The decline in numbers of small innovative businesses is, then, both a story of failure and of occasional success, with a few small enterprises surviving a difficult Darwinian selection of the fittest. The hope of the Russian government is that these survivors can both grow in output and be supplemented by better-equipped and more knowledgeable newcomers. It is still too early to know just how realistic this hope is.

Surveys of small innovative enterprises conducted in 1999, 2000, and 2003[32] with a sample of almost 500 showed that at the end of the 1990s the main obstacle to their development was inadequate financing, but undeveloped S&T infrastructure (including the low effectiveness of such organizations as Technology Parks and ITCs) was also a problem.

Almost 90 percent of Russian small innovative businesses work for the internal market; 44 percent work for both the internal and external markets and only 10 percent for foreign buyers. Often a gradual transition is made from the internal to the foreign market, but there are cases in which the movement was in the reverse direction—from the external to the internal. This reverse transition occurred in those instances when the purchasers of technologies in Russia were at first wary of the production of domestic small firms but later learned to choose good products whatever the source.

The initial capital for small innovative enterprises in Russia is most commonly the investment of funds by the entrepreneur-organizers of the firm. Less frequently the money comes from bank credits or independent venture capitalists (often called "business-angels"). Thus, the pool of money available for innovative enterprises is restricted, and these limits have not been overcome by the existence of such special funds as the Fund for Assistance or the RFTD, which have budgets far too small to meet the needs. Possibilities for further funds and for market expansion exist if Russian firms make partnerships with Western businesses or distributors, but many Russian enterprises prefer to work in the domestic market where they feel more confident and do not experience much competitive pressure.

At present, small innovative firms arise in Russia in one of three basic ways, which in shorthand we can call "research lab initiated," "trade initiated," and "business initiated." In the first case, scientists or engineers in a lab develop a product that they themselves try to commercialize. They are following a "technology push" model. In the second case scientists first leave their research organizations, get involved in trade and middleman activity, earn a little capital, and then try to create a small innovative enterprise. In the third case representatives of business study the needs of the market, develop R&D, and then move into manufacturing. These business people are following a "market pull" model.

Unfortunately in Russia the most common strategies are to follow the first and second approaches, even though the third is most often economically successful. Too frequently efforts to create innovative industries in Russia are based on a very weak understanding of the market. This is particularly true when the first model (research lab initiated) is followed, as is often the case.

Representatives of any one of these three approaches may or may not seek patent protection for their innovations. Such protection helps regularize and legalize high technology business, but many initiators of such enterprises in Russia are reluctant to seek such legal protection for their innovations, since they do not like to reveal technical details as required by patent application, and they, perhaps understandably, do not believe that Russia offers strong legal protection for business. As a result they often proceed unprotected, relying on what might be called trade secrets. Such a practice sometimes occurs in the West as well, particularly in very rapidly developing areas of technology (where obsolescence often outstrips the patent system), but the choice between protection and non-protection should be a carefully considered business decision, not one based on distrust of the legal system

The legal situation of small enterprises is complex and difficult. For example, currently government R&D organizations can be co-founders of private enterprises,[33] but they cannot possess founder shares, nor can they possess intellectual property in these new enterprises. To an American accustomed to a separation between government and private business this requirement may not seem onerous, but, as noted earlier, one should remember that almost all R&D organizations and higher educational institutions in Russia are government ones, and therefore there is relatively little economic incentive for them to try to establish private businesses.

Much of the old Soviet structure of the scientific establishment has been preserved. The most widespread and recognized mechanism to stimulate technology development is the awarding of various "statuses" to existing research establishments by naming them Technology Parks, Innovation Technology Centers, Federal Research Centers, Technological Villages, and so forth. Such status acquired through renaming brings with it preferential financial support by the government. Business interests understandably lobby for the selection by the government of their favorite institutions and projects. An open question is whether businesses are attracting government funds for projects they should finance on their own. Once conferred, such governmentally awarded status remains with the organization practically forever. Serious reexamination of the status in light of results is not conducted.

A crucial question for all governments interested in stimulating innovation activity is "What is best done by the government and what is best left to the market?" It is clear that when the Russian government gives status or money to certain centers or industries, calling them ITCs or Technology Parks, it is "picking winners" rather than allowing the market to do the choosing. Critics might say that such an approach is a vestige of the central planning mentality, and to a certain extent it may be. However, this issue is complex and subtle, and can not be resolved in a simplistic way. Governments in all industrial countries, including the United States despite its strong allegiance to the market principle, try to help innovative industry, operating through such government organizations as, in the United States, the National Science Foundation or the SBIR, and in France the ANVAR program. Just where to draw the line between the interplay of central government action and the free market is not easily defined. Both the government and the market seem to have important and necessary roles. In Russia, at a time of crisis in Russian science and technology, continued support from the government is especially needed. But since government support and control have such strong traditions in Russia, the "weaning away" process is particularly difficult. This is a problem with which Russia will be wrestling for many more years.

In the years since the collapse of the Soviet Union the Russian government has been trying to create the conditions necessary for the successful commercialization of scientific knowledge. It has established investment funds, local organizations providing support to start-up companies, and a new legal framework for patents and commercial activities. While considerable progress has been made in this effort to create an "innovation infrastructure," so far the actual products have been disappointingly small. The export of high technology is still a very modest part of the Russian economy (except in the military area), much less than the overall size of the economy and the underlying strength of the scientific community would seem to warrant.

This difficulty is an illustration of how formal structures (funds, organizations, laws) are not, in themselves, adequate for stimulating high technology, even though they are certainly necessary. Attitudinal changes are also needed, and these seem to come particularly difficultly in Russia. Very few investors in Russia are interested in high technology; they prefer to invest in the extractive industries and in retail trade, usually disposing of their investments in places where they have friends, family members, or acquaintances.

Those start-up and small technology firms that have been born in protected areas (Technology Parks, Special Economic Zones, etc.) seem reluctant to leave those areas, preferring to stay small in a comfortable place rather than emerging into the larger, more risky general economy. Profitability is sometimes less desirable than security.

We see, then, that Russia faces many obstacles in its effort to create a new knowledge-based economy. These obstacles include weak understanding of the market, a continued preference for uniform support to industries rather than competition among them, a scatter-shot and fragmentary approach to stimulating innovating industries, a weak legal basis protecting innovations and intellectual property, and inadequate linkages between science and business.

We should emphasize that near the end of the period of our study—in December 2006—the Russian government for the first time placed rather large sums of money (over $600 million U.S.) in organizations designed to stimulate commercial activities in high technology. The two organizations created at that time—the Russian Venture Company and the Russian Investors Fund for Information-Communication Technologies—deserve future scrutiny. It is far too early to judge their success.

As we will indicate in our final chapter, some of Russia's problems in high technology are specific (the size of the country, the continuing legacy of Soviet practices) and some of them are common to many countries (the effort to minimize market failures and government failures in stimulating innovation). Countries everywhere also wrestle with these questions.

6   International Support of Russian Science: History and Evolution

The History of Programs for Help and Cooperation in Science

During the last fifteen years the former Soviet Union (FSU) has been the object of the largest international scientific aid program in history. It is impossible to calculate the total amount of foreign money donated, invested, or co-financed by foreign organizations in FSU science during these years, but just three organizations—the ISF,[1] the ISTC,[2] and INTAS[3]—have put well over a billion U.S. dollars into Russian science and technology in this period. And dozens of other organizations have also made their funds available to Russian scientists and engineers.

The year 1992–1993 can be considered the date of the beginning of the activity of foreign foundations and international organizations in Russian science; in this year a number of them opened offices in Russia and announced grant competitions. However, aid programs administered from abroad began as early as 1991, not long before the disintegration of the USSR. Even earlier there were individual initiatives that are best described not as aid programs but as academic exchanges. The latter were based on agreements with the Academy of Sciences of the USSR and a few Soviet universities and were conducted through such foreign organizations as the National Academy of Sciences in the United States, the International Research and Exchanges Board (IREX), also in the United States, the German Academic Exchange Service (DAAD), the Royal Swedish Academy of Sciences, the British Council, and others.

If one goes back further in history, the foreign support of Soviet science actually begins from the moment of the end of the Civil War in Soviet Russia in the early 1920s, when the Rockefeller Foundation was giving grants to young researchers all over the world (as a rule, to researchers younger than thirty-five). These grants, usually for study abroad, were awarded to citizens of many countries, including forty-eight Soviet scientists. The list of recipients reads today almost like a "Who's Who" of early Soviet science. Lev Landau, a future Nobelist in physics, and Nikolai Luzin, often called the founder of the famous Moscow School of Mathematics, received Rockefeller stipends. In the early thirties the political situation in the USSR became very strained, and travel abroad became practically impossible; fourteen additional Rockefeller grants were unclaimed by Soviet recipients unable to travel.[4] One of the results of the work of the Rockefeller Foundation was that the USSR became strong in genetics, an achievement that was later almost nullified by the rise of Lysenkoism, a false doctrine that denied the existence of the gene.

The second period of interaction of Soviet science and foreign organizations was in 1958–1971, when the dominant form was scientific exchanges. One of the chief announced goals for these exchanges was an increase in mutual understanding among the USSR and other countries. In 1959 the first inter-academy exchanges in the natural sciences began between the National Academy of Sciences of the United States and the Academy of Sciences of the USSR.[5]

The third period (1973–1991) was characterized by the development of intergovernmental S&T programs, often ones initiated for political or social reasons, such as the mutual desire to stop environmental degradation.[6] The number of participants in these programs and the level of financing significantly exceeded those of the earlier established (and still continuing) scientific exchanges. In 1977 the National Academy of Sciences in the United States published a report evaluating these latter exchanges, which stated that several scientific fields in the Soviet Union, especially physics and mathematics, were equal in quality to the best in the world, and that a few sub-fields surpassed the others.[7]

After the disintegration of the USSR, the fourth period of support of Russian science began, and it still continues, although in changing forms. The chief reason for this new assistance was the crisis in Russian science caused by the rapid curtailment of Russian government funds for science at a time when non-government funds for science simply did not exist. As a result foreign aid began to appear, at first from foreign private philanthropic organizations and later from intergovernmental agreements. At the same time, the intergovernmental S&T programs between the United States and Russia, initiated during the Soviet period, have continued to the present. On January 13, 2006, the two countries renewed the agreement for ten more years. The new agreement emphasized applied science in such areas as ecology, health care, and energy, as well as anti-terrorism technology.[8]

The foreign organizations that began activities in Russia in the nineties brought with them methods of selecting grant recipients and financing projects that were different from those in the Russian experience. In the Soviet period all programs had been strictly regulated by the centralized government, but now, after the disappearance of the Soviet Union, private and local initiatives began to appear, each with its own goals and priorities. The competitive principle was introduced into grant programs, and there was much greater freedom of choice of topics by applicants, as well as freedom of selection of grant recipients. Even organizations that had earlier been active in the USSR, such as IREX and Fulbright, took the new conditions into account and after 1992 changed the basic principles of their work in Russia.

Attitudes in the United States toward Aid to Russian Science

Private Western philanthropic organizations became active in Russian science before Western governments responded. Private organizations frequently act more quickly than government agencies because they do not need to work through such large bureaucracies. Furthermore, they often have a broader definition of "mutual benefit" than do governments.

However, fairly early—in May 1992—a report on the subject of Russian science was prepared for Alan Bromley, the U.S. president's advisor on science and technology. Titled "The Reorientation of Research Potential of the Former Soviet Union,"[9] the report set up two long-term goals for the United States with reference to Russian science: (1) The prevention of brain drain (especially to "rogue nations"), and (2) the formation of a democratic society with a market economy. The report emphasized that a program of aid to Russia must be constructed on a mutually beneficial basis and not be a simple assistance program for scientists and scientific institutions in the former Soviet Union. The 120 specialists who prepared the report recommended special attention to the reorientation of Russian scientists away from military tasks; they also asked for attention to researchers in physics, chemistry, biology, geology, and mathematics. Since about 75 percent of Soviet scientists were engaged in military R&D, their dispersal alarmed Western government leaders, who feared that they might find employment with rogue regimes and reveal the military secrets they possessed.

The authors of the report proposed "mutually beneficial" joint research projects with Russia, and also "goal-oriented" projects. The latter would utilize and preserve unique laboratories in the FSU, as well as information possessed by Russian researchers that was valuable for U.S. scientists. They proposed that financial assistance be given alike to individual researchers, scientific groups, and unique scientific facilities.

Many international organizations and foundations—not just in the United States—set up similar goals: the stemming of brain drain, the conversion of military research to civilian research, and the promotion of research that was mutually beneficial. A difference between private foundations and government initiatives was that private foundations such as George Soros's International Science Foundation and the John D. and Catherine T. MacArthur Foundation often emphasized philanthropic, charitable, and socially beneficial goals, such as aid to needy scientists and their families, environmental improvement, human rights, health problems, the judicial system, education, public media, and political reform. Similarly, the ISF (International Science Foundation) of George Soros was not concerned directly with military science but rather with alleviating the economic situation of scientists in the former Soviet Union. The private foundations were usually not insistent that their programs be mutually beneficial, at least not in an obvious way. The government programs tended to be more instrumentally focused on preventing weapons scientists from emigrating abroad to rogue nations, converting military research to civilian goals, and mutual benefit.

However, both private and government officials usually saw support for democracy and market reforms as mutually beneficial. In 1993 President Clinton of the United States emphasized three basic goals of U.S. government aid to Russia: supporting democracy, supporting market reforms, and supporting the effort to diminish the nuclear threat.

Concrete steps for the attainment of these goals were embodied in the Nunn-Lugar program, which included appropriations for the support of Russian science. This program was financed by the U.S. Department of Defense and was intended to assist the conversion of military industry in the FSU to civilian purposes. On the basis of funds from this program several international initiatives were launched, and in 1995 such funds, together with a $5 million contribution from George Soros, were used to create the Civilian Research and Development Foundation (CRDF). The CRDF is a nonprofit organization authorized by the U.S. Congress and established in 1995 by the National Science Foundation. It is an unusual public-private partnership promoting international scientific and technical collaboration, primarily between the United States and Eurasia, through grants, technical resources, and training. It emphasizes conversion of military research to civilian purposes in many of its programs, but also promotes education and market reforms.

An act of the U.S. Congress that regulated the distribution of government funds supporting activities in Russia (including science) was the "Freedom for Russia and Emerging Eurasian Democracies, and Open Markets Support Act of 1992" or FSA. The FSA stipulated thirteen priorities for aid, of which several were directly connected to science, such as the improvement of the environment and quality of life, the security of nuclear reactors, the development of telecommunications, and the regulation of migration. A whole section of the FSA was devoted to the problem of the conversion and nonproliferation of weapons. The FSA also stated preferences for recipients of support, such as scientists outside the capital cities of Moscow and St. Petersburg (what is often called regional science), youth (undergraduate and graduate students, young scientists), and women researchers.

Private U.S. foundations, for their own reasons, often favored similar recipients. Even today the majority of U.S organizations active in Russia, as will be shown below, try to take these priorities into account in their work.

The peak in political attention to Russia lasted until 1995, after which financing for such programs, both government and private, began to diminish. In 1996 several programs were closed down (for example, the program of joint research in IREX) and many others were scaled down (for example, the U.S. National Science Foundation after 1995 supported fewer joint projects with Russia). With time several initiatives were curtailed. The International Science Foundation, created by George Soros, ceased its activities in 1996, not because of changing U.S. government policy but because Soros shifted his priorities in philanthropic aid. Support for Russian science did not end, however. In the middle nineties several private U.S. foundations even developed additional programs.

The European Union: Priorities in Cooperation with Russia

In the European Union political motivations for supporting Russian science were less broad than in the United States. The European countries basically pursued the goals of mutually beneficial scientific and technical cooperation. Their main areas of interest were those lines of research that were important scientifically and economically for their own countries and were highly developed in Russia; the European countries did not try to raise the level of "backward" S&T fields in Russia.

However, one political precept important in the European efforts was the concept of mobility. In the opinion of West European specialists, scientists who work in more than one laboratory during their careers enrich their knowledge and experience and thereby improve their professional level. Furthermore, this mobility implies new contacts, the development of cooperation, and, finally, new possibilities for technological development.

The policy of encouraging mobility and the development of partnerships was embodied in the Program of INTAS, created in June 1993. This policy was borrowed from an earlier European program, Human Capital Mobility. According to the regulations of INTAS, each project must involve two partners from the countries of the European Union, and in that way stimulate contacts between Russian and European scientists. In the opinion of the European partners, mobility is also a way to limit brain drain from Russia, since it means that scientists can go back and forth at will.

When choosing areas of cooperation INTAS relied on criteria such as the quality of scientific research in Russia in the given area, the potential for commercialization, and also the contribution of the project to political stability and sustainable development. In France a priority was cooperation in fundamental physics and mathematics, and later in biology and biotechnology. The European countries recognized the high level of development of Russian science, which in some areas exceeded that of Western Europe.

Today, after more than fifteen years of the work of foreign organizations with Russian science, we can evaluate the evolution of their policies, the results of their work, and the effectiveness of their programs.

A Typology of Foreign Organizations and Foundations in Russian Science

We can identify several different types of foreign organizations and foundations working in Russian science and technology, distinguishing them on the basis of their organizational-legal form and the sources of their funds.

In organizational and legal terms some of the organizations are foundations (John D. and Catherine T. MacArthur Foundation, Ford Foundation, Carnegie Corporation of New York), some are associations (INTAS, ISTC), and some are public-private partnerships (CRDF).

If we classify the organizations in terms of the source of their funds, the foundations can be separated into those having their own endowments (for examples, those established with bequests from private individuals, such as Ford, Rockefeller, Carnegie and Soros) and those that receive their funds from other foundations or the government and serve as middlemen. For example, IREX receives funds from private sources (Carnegie Corporation, private universities) and also from the federal government (U.S. Agency for International Development, or AID). Such organizations receiving money from elsewhere are sometimes called fund-driven. Finally, there are organizations funded exclusively by governments (Fulbright Program, Netherlands Organization for Scientific Research, the German Research Society or DFG, the U. S. National Science Foundation, and the like).

The majority of foreign organizations actively working in Russia are financed from government funds or have mixed funding. In the beginning some of the organizations distributed government funds exclusively, but then, when government financing diminished, they sought additional sources and developed partnership ties with private organizations and/or with government agencies in other countries. The share of private funds in such organizations and foundations is very small—usually not more than 10 percent, although with time that share will probably grow. Examples of such an evolution are INTAS, the Max Planck Society, and the German Academic Exchange Service (DAAD).

In terms of country of origin, the programs best known in Russia are those of the United States, of the European Union, and also of international organizations representing the interests of several different countries (first of all, the International Science and Technology Center or ISTC, with its very large budget). There is also a large number of foundations and organizations that have no offices in Russia but nonetheless have international programs supporting scientific research in which Russian scientists can participate, along with those of other nations. (An example is the Howard Hughes Medical Institute.) Of the foreign foundations and organizations active in S&T in Russia, according to one listing, 34.3 percent are from the United States, 14.7 percent are British, 10.7 percent are from the European Union, 10.1 percent are German, 5.8 percent are Japanese, and 4.1 percent are French.[10]

The foundations distribute funds on the basis of competitive grants and special programs and initiatives. There are also "contract works," in which the foundation fulfills specific tasks on behalf of a government agency.[11] The most common form of financing is individual or group grants, given directly to Russian scientists to conduct research in Russia or abroad, and also joint projects in which foreign and Russian partners participate together. Special projects include those aimed toward commercialization of innovations, often jointly financed.

Increasingly, foundations launch programs on the basis of shared financing in which Russian government and regional sources contribute funds. Such projects are supported by the MacArthur Foundation, the Carnegie Corporation, the Ford Foundation, CRDF, and NWO (the Netherlands Organization for Scientific Research). One and the same foundation or organization may support several types of programs. Diversification in the activities of the foundations is becoming stronger.

The majority of foundations working in Russia finance research connected with social issues (the social sciences, the humanities, and also ecology, medicine, and agriculture). Support for research in the natural sciences and for the commercialization of R&D is currently given primarily as joint projects (in the programs of INTAS, ISTC, CRDF, NWO, and 6FP, the Sixth Framework Program of the European Union).

The International Science Foundation (ISF): Emergency Aid and an Important Prototype

Despite the fact that the ISF was active in Russia for only four years (1993–1996), it played a significant role, both for what it did in that short time and as an example for others, including Russian organizations. The money given out by the ISF during these four years remains the largest foreign contribution in the support of Russian science within such a short time period. The contribution and influence of the ISF are connected with the large scale of its work and also with the fact that it was the first foreign organization that demonstrated the legal and organizational principles of a science foundation— principles that would later be very important for Russia.

From 1993 until 1996 the ISF invested almost $130 million in basic science in the former Soviet Union (see table 6.1), of which $95 million were given out through competitive grant financing (a new mechanism for Russia). In all the programs of the ISF, approximately 50,000 scientists of the FSU received support. In 1994–1995 funds from the ISF amounted to about 13 percent of the total expenditures on basic science in Russia. The chief goals of the ISF were to support the best research groups at a time of great need and to disseminate new principles of financing and managing scientific projects.[12]

As table 6.1 shows, the ISF supported a number of different activities, but it emphasized three particular programs: emergency aid, long-term grants, and travel grants. All three of these forms of aid came to Russian science at a time when money was very scarce (and for activities such as travel, practically nonexistent).[13]

Although the program for emergency aid provided less than one-fifth that disbursed for long-term grants, it has a special place in the memory of Russian scientists. In 1993 and 1994, when salaries for Russian scientists simply disappeared, George Soros pushed through this program with striking rapidity. From the moment of its announcement to the completion of grant distribution, less than four months passed. Under this program 20,763 scientists in Russia received grants of $500 each. Although this sum may not seem large, it was, at the existing exchange rate, almost a year's salary for most Russian scientists (perhaps an irrelevant fact, since most of them were not receiving even these miserable salaries). Finding food for a scientist's family to get through the winter was certainly possible with $500.

The emergency aid program was followed by a program of long-term grants with full peer review that served as a model for other foundations and the Russians themselves. Under the long-term grants program 15,000 scientists of the FSU received awards that supported them for at least one year of research. Such

Table 6.1. Financing of various ISF programs (statistics for the entire former USSR)

Name of program	Amount of financing in financial year in millions of dollars				Total expenditures for 4 years in millions of dollars
	1993	1994	1995	1996	
Expenditures in all ISF programs	16.937	46.682	54.087	9.081	126.787
Specific programs of above:					
Emergency aid	12.174	3.109	0.302	—	15.585
Long-term grants	0.224	29.534	44.433	5.880	80.072
Travel grants to attend scientific conferences	3.142	7.559	2.300	1.366	14.367
Library support	0.656	1.425	1.460	1.236	4.777
Telecommunications support	0.490	0.874	2.777	—	4.141

Source: database of PHRI (Public Health Research Institute), New York office of ISF, 1998.

grants were awarded in approximately one-third of all scientific-research organizations and universities in Russia.

Recognizing that lack of travel money was causing scientists in Russia to lose their contacts with Western scientists (other than by slow international mail), the ISF also promoted an ambitious travel grant program under which approximately 10,000 scientists from the former Soviet Union were able to attend more than 2,000 international science conferences. With a similar goal of improving international communication in science, the ISF promoted a telecommunications program providing free access to the internet to many scientific organizations in Russia. The ISF also created fiber-optic telecommunication networks in local regions. Other funds went to 500 libraries for subscriptions to 130 science journals.

The program of long-term grants for research was the largest initiative of the ISF, making up 63 percent of the total budget. This program from the beginning was assigned a special significance: one of the chief goals of the ISF was promoting new approaches to the organization and financing of scientific research. An important characteristic of this program was its basis in a classical Western model for the distribution of grants by means of an open competition and the evaluation of proposals by international experts within a peer review system. International experts, working on such peer review committees for the ISF, voluntarily contributed the equivalent of 22.5 work-years for the evaluation of projects presented by scientists in the FSU. The principles of organization and managements used in this program had a long-term significance for Russian science. In the future, several of its approaches would be adapted by Russian government science foundations.

The chief innovations that would help Russia in its development of a new system of organizing and financing research were: (1) Direct support to scientists, and not to organizations, by means of a system of competitive grants, with the chief criterion for award of grants being the quality of proposals being submitted; and (2) introduction of peer review with the participation of international experts—that is, evaluation by equals. The peer review was independent, anonymous, and, in most cases, conducted by experts from a country different from that of the applicants. The experience of the NSF and the NIH in the United States provided models for this peer review system. The RFBR, created in 1992, already operated through a system of open competitive grant financing, but the ISF had the additional specific characteristic of international peer review. The fact that the evaluation of projects was done primarily by foreign experts gave Russian scientists the feeling that the competition was impartially and fairly conducted.

In addition to the innovations connected with the program of long-term grants, the ISF used a whole series of approaches that were important for the development of new principles of management in Russian science. Among them were flexibility in the structure of management, the interrelatedness of its programs, and the development of a system of public relations.

The ISF regarded science as a complex, multi-component system, and therefore simultaneously supported individual scientists, scientific institutions, and purchases of equipment and instruments. Great attention was given to modernization of equipment. Such purchases could be made within the framework of the "first aid" program, the long-term grants, and in supplementary grants sometimes made outside the competitive framework. The purchase of personal computers for work on projects was accompanied, as a rule, by the establishment of e-mail and use of the internet.

Because of its activities, the ISF became well known in Russia in a very short period of time. According to a survey made in 1994–1995 in scientific organizations of various regions of Russia,[14] scientists were more familiar with the ISF than any other foreign foundation. In the majority of the areas outside the capital cities of Russia, scientists knew only about the ISF at a time when scientists in Moscow could name more than thirty foreign foundations they had heard about one way or another.

The significance of the ISF consisted in the fact that in its programs there was a learning process for all participants. Thus, librarians learned to work with foreign publishers; Russian scientists serving as peer reviewers became accustomed to evaluating projects while observing international norms; scientific researchers applying through the short-term and long-term grant programs learned how to write good proposals and later prepare works for publication. Moreover, scientists were able to evaluate themselves by comparing their research with similar work going on abroad. The system of grant financing of the

ISF was approved by 75 percent of scientists surveyed in 1993–1994 and almost 86 percent of those surveyed in 1995.[15]

In interviews conducted two years after the completion of their ISF grants, scientists in Russia were asked, "Was the long-term grants program simply a method of keeping individual scientists and institutions alive, or did it promote new scientific results?" Their answers differed by field and by geographic location of their institutions. The majority of mathematicians replied that the ISF grants permitted them to develop new scientific ideas. Usually mathematicians did not need expensive equipment, only free time—something the grants from the ISF gave them. Those scientists who were working in areas of research where expensive equipment and field work were necessary were less positive, noting that conducting a full expedition was not possible on the basis of an ISF grant. Nonetheless, for some institutes, even those needing expensive equipment, grants from the ISF were critically important. For example, in the Ioffe Physico-Technical Institute in St. Petersburg the grants from the ISF were for a while equal to about half of the annual budget (normally supplied by the Russian Academy of Sciences).[16]

The introduction of a Western model of financing science provided not only new mechanisms for distributing funds but also a new ethic of relations among scientists. Russians began to understand such concepts as conflict of interests, as well as the maintenance of confidentiality. The Western approach even introduced changes in definitions of research problems and how to solve them. The Russian researchers noticed that the Western funders and peer reviewers believed that research even in basic science should result in some kind of concrete finding, and not be merely open-ended theoretical inquiry. Russian researchers who in their proposals to Western funders demonstrated attention to possible applications of their work were often pleased to find that they succeeded in competition with "classical" theoreticians. Thus, the Western model even influenced the intellectual content of proposals, a fact that deserves additional study, since it is possible that there may be both gains and losses in such influence.

The ISF also helped achieve new economic effects. Thanks to its efforts, payments in foreign currency in the former Soviet Union were legalized, at least in science. The ISF was one of the first organizations to introduce scientists to the possibility of finding funding alternatives, in both native and foreign currencies. This was a contribution to the forming of a new post-Soviet model of science. The new grants gave an internal impulse to research and a stimulus for seeking out partners and new sources of financing.

In this way, the ISF first introduced into Russia ideas about the organization and funding of science that today are commonly accepted as normal: the concept of grants, the peer review system, international evaluation of projects, goal-oriented support of scientists instead of organizations as a whole, and the use of new principles of grant management. The introduction of the grant system meant not only the appearance of a new source and form for the support of science, but also the birth of a new mentality about research. All these elements were a contribution to the formation of a market-oriented culture in which science could function.

International Science and Technology Center (ISTC)

The idea for creating the ISTC was born in 1992, when the ministers and secretaries of foreign affairs of Russia, Germany, and the United States agreed to finance research aimed at conversion from military to civilian goals.[17] At that time they specified what topics and areas would be the object of the joint initiative. Russian research institutes that had been conducting high-tech defense work would be supported for conversion purposes. Emphasis was on nuclear, chemical, and biological weapons and means for their delivery.

Working out the details and achieving full agreement required two more years, and in November 1994 the ISTC was established. The ISTC is an intergovernmental organization that currently includes representatives of the European Union, Switzerland, the United States, Canada, Japan, Norway, the Republic of Korea, and Russia.

The ISTC has three goals: reorientation toward civilian R&D of scientists from the FSU who have knowledge and skills in the area of weapons of mass destruction and their delivery; support of fundamental and applied research that assists in solving national and international technical problems; and support for the transition to a market economy and the integration of former weapons scientists in the world scientific community.

The ISTC has a very large budget and has been a major international influence on Russian science. In 2005 alone its budget for new projects was over $51,000,000.[18] Total ISTC project funding in all categories surpasses $725,000,000.[19] The activities and impact of the ISTC will be discussed in more detail in following sections.

International Association for the Promotion of Cooperation with Scientists from the Independent States of the Former Soviet Union (INTAS)

The INTAS association was registered in 1993 as an international noncommercial organization, in conformity with the laws of Belgium. According to Belgian law, however, such an association or foundation must be a private organization. As a result, from a legal standpoint INTAS is a private foundation, but with primarily government financing. The founders of INTAS were fifteen members of the European Union (EU), and also Norway and Switzerland. At present thirty-three nations are members of INTAS. More than 90 percent of the budget comes from the European Commission (EC), and the rest from private sources. Beginning in 1995 the European Commission has headed the Assembly of INTAS. This Commission has veto rights on any decision of the Association. In this way the state interests of the member countries of the European Union have a great significance in such decisions.

The goals of INTAS are helping scientists of the former Soviet Union and promoting cooperation between scientists of the former Soviet Union and the countries of Western Europe in R&D. Between 1993 and 2004, INTAS funded approximately 4,000 projects and activities in the FSU with a total budget of over €200,000,000. With this large budget INTAS has been a major influence on science in the FSU. Its activities and impact will also be discussed in following sections.

In September 2006, the European Commission agreed to discontinue INTAS as of January 1, 2007. No new activities would be initiated by INTAS after that date, although ongoing projects would be continued until the dates stated in the original grant agreements.

The Evolution of Goals and Priorities of Foreign Organizations and Foundations

The goals of programs and foundations utilizing government funds were closely connected with the state interests of foreign participant countries, and also with their science policies. Therefore, the changes occurring over the last fifteen years in Russia and in the donor countries influenced the goals and directions of activity of foreign and international organizations and foundations. The goals of programs relying on private funds were more independent from state interests, but even they were affected by the evolution of state policies (such as the Russian government's attitude toward outside funders of all types).

Foreign organizations and foundations initiated programs in Russian science for somewhat varying reasons (table 6.2), of which the most common were four: (1) the desire to help Russian scientists who were in distress because of the financial crisis after the breakup of the Soviet Union; (2) the hope for mutually beneficial cooperation between scientists in Russia and the donor countries; (3) the hope for better mutual understanding between Russia and the donor countries; and (4) worry about the export of knowledge of weapons of mass destruction (security concerns). Later evidence on relative amounts of money spent on different topics shows that the last reason, fear of an outflow of Russian scientists to "unfriendly countries," was of considerable significance, but it was actually named by only three organizations—ISTC, INTAS, and CRDF.

The announced goals of foreign organizations and foundations were not always identical with the ways they began to work in Russia. For most foundations the main motivation was not philanthropic aid to Russian science but the development of a mutually advantageous partnership and cooperation (table 6.3).

In the initial period of the work of foundations, basic attention was focused on philanthropic, partnership, and security functions, but later the function of reforming science in Russia became important. The commercial function also grew in strength. One and the same foundation often combined several functions, as tables 6.1–6.3 show. For example, both the ISTC and the CRDF served a number of functions.

Table 6.2. Announced goals of foreign organizations
and foundations supporting Russian science and technology

Reasons that foreign organizations began to support Russian science	Foundations and organizations initiating programs for the given reason	Functions the organizations and foundations fulfilled
The crisis situation in Russian science	NSF, ISF, INTAS, ISTC, CRDF, NWO	Philanthropic
The development of mutually beneficial cooperation	INTAS, Wellcome Trust, German Research Foundation (DFG), British Council, von Humboldt Foundation, Max Planck Society	Partnership relationship
The development of understanding between Russia and other countries	MacArthur Foundation, Fulbright Program, Royal Society (Great Britain), von Humboldt Foundation	Philanthropic and geopolitical
Prevention of emigration of Russian scientists to "unfriendly countries"	ISTC, INTAS, CRDF	Security-oriented
Fostering democracy in Russia, promoting civil society	IREX, Ford	Reform-oriented; geopolitical
Promoting the German language	DAAD (German Academic Exchange Service)	Cultural, geopolitical

("Security-oriented" here means related to nonproliferation of weapons; "geopolitical" here means increasing the political influence of the donor country.)
Source: I. Dezhina, interviews, 2003–2007.

Selection and Peer Review in the Foundations; Participation of Russians in Decision Making

The priorities of the foundations are to a certain extent reflected in their mechanisms of selecting recipients and peer review, despite the fact that these processes are ideally independent and impartial. An analysis of the activities of foundations shows that there is great variety among methods for making evaluations, approaches to peer review, and procedures for taking decisions about funding. Understanding these details is important in responding to criticisms of the programs. Critics, particularly in Russia, sometimes assume that awards were often made with hidden goals and on the basis of biased selections. An examination of these selection procedures also reveals the complexity of the situation in which foreign foundations find themselves when they make grants on the territory of a formerly hostile and still somewhat suspicious country.

At the same time there are certain general characteristics, and also general

Table 6.3. Actual functions of foreign organizations and foundations in Russian S&T

Goals of foreign organizations and foundations	Foundations and organizations pursuing these goals	Functions performed by the organizations and foundations
The development of mutually beneficial cooperation	NWO, INTAS, Wellcome Trust, DFG, British Council, von Humboldt Foundation, Max Planck Society, Fulbright Program, ISTC, CRDF	Partnership relationships Commercial relationships
Support of the best scientists and strongest scientific developments (in the initial stage—aid to scientists in difficult situations)	NSF, ISF, NWO, Wellcome Trust, DAAD, von Humboldt Foundation, INTAS, ISTC, CRDF, Howard Hughes Medical Institute	Philanthropic, scientific
Strengthening the existing and forming new S&T infrastructures	MacArthur Foundation, CRDF	Reform-oriented
Promoting democracy and human rights in Russia, building a civil society	MacArthur Foundation, IREX	Reform-oriented
Redirecting scientists in the former defense complex toward civilian research	ISTC, CRDF	Reform-oriented Security-oriented
Helping the transition to a market economy	ISTC, CRDF	Reform-oriented Commercial
Giving foreign scientists access to scientific equipment and information in Russia	NSF, ISF, Wellcome Trust	Scientific, partnership relationships

Source: I. Dezhina, interviews, 2003–2007.

problems of organizing and conducting evaluations by foreign foundations operating in the former Soviet Union.

- The first general characteristic of selection is the obligatory presence at one stage or another (usually the first stage) of an independent and anonymous evaluation by specialists of the grant proposal, graded by a voting system (peer review system).

    Peer review is the system of evaluation most accepted in the scientific world, despite the fact that this method, like any other, has some flaws. The peer review system is not entirely free from conflict of interest, since the peer reviewers are often themselves seekers of grants from foundations. Furthermore, this is an "inertial" system and therefore may be too conservative, supporting already well-known directions of research. Several methods exist to diminish the possible negative features of the peer review system.[20]

One way to diminish negative features is to change the scheme for applying for a grant so that financial and accompanying materials are presented only after the proposal has passed through the first peer-review evaluation and has become a potential candidate for support. This approach is being used by a growing number of organizations and foundations. Applying to foundations is often time-consuming and cumbersome, and can be very inefficient for researchers if the success rate is low.

In Soviet times evaluation was usually performed by the relevant department or organization, but peer review per se was practically nonexistent. Competitions for carrying out government S&T programs were considered open, but in fact they were yet another form of support to institutions as a whole and not to individual working groups.

- A second typical element of evaluation is a two-stage procedure. Usually this is a combination of anonymous evaluation with a discussion of proposals in specialist committees. This multi-stage evaluation reduces subjectivity in grant making and helps to overcome the flaws of the peer-review system.
- A third general characteristic of the procedure of evaluation is that the final decision about awarding grants is taken, as a rule, by specialists, in the meetings of the expert councils. However, in several foundations the final decision is taken by administrators (for example, in the ISTC, CRDF, and INTAS).
- A fourth characteristic is that the experts are, as a rule, foreign scientists. They participate at all stages of the selection of projects. In joint research foreign evaluation is obligatory—since foreign experts evaluate the significance of the theme of the Russian partner for their country.
- A fifth characteristic of selection is that in financing scientific projects the decisive factors are the scientific quality, novelty, and significance of the projects. Choice of this or that project depends on a set of criteria or priorities that each sponsor country establishes. European partners do not emphasize exactly the same criteria as U.S. or Japanese, and the reverse. Special preferences (such as the support of women, young scientists, geographic distribution of researchers) are taken into account, as a rule, in cases in which it is necessary to make choices among or between projects that in other respects are equal. An examination of the list of special categories shows that the majority of foundations give special attention to the support of young scientists. A much smaller number of foundations give attention to geographic distribution, and still fewer to women researchers.

The organizations that distribute the most money in Russia, and therefore most strongly demonstrate foreign practices of grant administration, deserve special attention. These, without question, are the international organizations ISTC and INTAS. The process of evaluation in CRDF also has interest; CRDF developed and adapted approaches used by NSF in the United States.

Evaluation of Projects by the ISTC

The ISTC has two types of programs:

(1) Scientific projects financed by the foreign participant countries of the ISTC and chosen by three parties: the United States, Japan, and the European Union. Each has its own peer review committees, and they independently decide if they will finance a given project.

(2) Partnership projects: Commercial companies can support a project through the ISTC, contributing an additional 5 percent for ISTC administrative expenses.

Because several countries participate in the financing of projects in the ISTC, the procedure of evaluation is multi-staged and rather complicated. The partners must make decisions not only about the quality of the project but also about who will finance it and to what degree.

The main motivation in ISTC programs is to prevent brain drain and reorient scientists in the defense sector toward civilian R&D; this motivation influences the chief criteria of choice. Therefore the first criterion—the quality of the project—joins with two others—the location of the institution, and the previous concerns of the leader on the Russian side. Closed, specialized cities (centers of classified research dating from the Soviet period) have an advantage, and the most successful scientists are those in the defense sector who also have experience in civilian research. Recently another priority has appeared: the potential for commercialization of the project, which the ISTC helps facilitate. Other special preferences (regional diversification, support of young scientists or women scientists) do not exist in ISTC programs.

In submitting the proposal to the ISTC the applicant(s) must have the permission of the relevant ministry, confirming in this way that the proposal does not touch on government secrets. Proposals are at first evaluated by specialists on the Scientific Advisory Committee, on which representatives of different countries sit. Each country conducts its own evaluation. In cases in which there are difficulties in evaluation, the proposal is sent out for supplementary peer review. In the United States, for example, such a proposal is sent for peer review to no less than three experts from national laboratories or from the NSF. Then the reviews go to the Scientific Advisory Committee, where each project is judged and given a preliminary evaluation. The evaluation scale has three levels.

Projects that are provisionally selected go to the Coordination Committee.

In its meeting there is a discussion of the priorities of projects by each side, their reconciliation, and a decision about which side will finance the project. Sometimes projects are financed by only one of the partners—for example, by the United States or Japan. The experience of evaluating projects in the ISTC has shown that Japan prefers to finance projects separately, since in this way distribution of rights to intellectual property is simplified. Then the applications go to the Governing Council, which makes the final decision.

A distinguishing characteristic of the ISTC is interim evaluation, i.e., the monitoring of the projects after they are under way. Periodic and special evaluations are conducted, including visits to institutions. Quarterly reports are required. Based on these evaluations and reports, recommendations are made for improving the projects.

Evaluations of Proposals in INTAS

In the evaluation of proposals in INTAS the staffs of the Association do not, to any practical extent, participate in the making of decisions; the selection of projects for financing depends almost entirely on the results of anonymous peer review by independent reviewers.

At the first stage each proposal is sent by email to three reviewers in the appropriate field of research, chosen on a random basis from a pool of INTAS experts. In this process the principle of regional diversification is observed: one must be from a European country, one from an Eastern country (for example, Japan), and the third from someplace else. In practice each INTAS expert evaluates on average from three to ten proposals. All work is conducted through the internet. If the experts differ strongly in their evaluations, then their commentaries, without the specific votes, are sent to coordinators in INTAS for a special decision.

Subsequently a ranked list of all proposals is drawn up and, as a rule, the top third receives financing without any further discussion. Then the top half of the proposals from the middle third are discussed at a meeting of the Scientific Council, the members of which are named by the General Assembly of INTAS. It includes both representatives of the European Union and the CIS. Specialists may be consulted, but only for those proposals on which the opinions of the anonymous experts differ. Members of the Council may decide to ignore one of the evaluations and in that way lower or raise the summary vote on the proposal.

The positive characteristic of the given system of evaluation is that any lobbying of the results is made difficult by virtue of the full independence of the anonymous experts. On the other hand such a system, like all such systems, contains flaws, especially the possibility that the anonymous experts are not the best qualified persons to evaluate the proposal. In some specialized fields INTAS has only a limited number of reviewers in its pool.

In selecting innovation projects there is a slightly different procedure: the evaluation is conducted by a special expert Council, which includes representatives of the Russian side (at present the interests of the Russian side are represented by the Fund for Assistance to Small Innovative Enterprises).

CRDF Proposal Selection

The method of selecting proposals in CRDF combines the approaches of both the U.S. NSF and also its ancestor in Russia—the ISF. As in the case of the NSF, the foundation employs a procedure of review by mail. Each proposal is evaluated by at least three reviewers, all U.S. scientists. The similarity to the ISF consists in the use of expert councils for each of the specialties. CRDF also uses requirements worked out by ISF according to which leaders of joint Russian-U.S. projects must possess the degree of Ph.D. (or "kandidat" or "doktor" in the Russian case) and have at least five publications in peer-reviewed journals. (If the degree was received no more than six years earlier, the number of publications in peer reviewed journals can be three.) At the same time the CRDF has intentionally not used the ISF approach to selecting reviewers in which the applicant himself or herself named six reviewers—no less than four foreigners and no more than two from his or her own country. Such an approach, in the opinion of the leadership of CRDF, led to a loss of confidentiality and excessive pressure on experts from the applicants. Instead, CRDF introduced a multi-stage scheme of selecting projects.

At the first stage, proposals not meeting the formal eligibility requirements are eliminated in the foundation office. At the second stage the proposal is given to an appropriate Expert Council, where it is evaluated by two members of the Council. The members of the Expert Council are all U.S. scientists. At the third stage the proposal is discussed at a meeting of the Expert Council, which consists of those experts who conducted the initial evaluation.

The further movement of the proposal depends on what summary evaluation it received and, correspondingly, what place it occupied in the general list of proposals in the given specialty. After supplementary evaluation the proposal comes to the second meeting of the Expert Council, where its fate is decided. In such a case, criteria of "the second order" may come into play, such as the presence on the project of female researchers, or young scientists (undergraduates or graduate students), or researchers outside major cities like Moscow and St. Petersburg. The final decision about financing, as in the case of NSF, is taken by staff of the Foundation, but the evaluation of the specialists is rarely overturned.

The described procedure concerns the joint research program. For other programs of the CRDF somewhat different schemes of selection are used (for example, in the evaluation of commercialization projects European scientists may be involved, or if co-financing of the program is envisioned, the evaluation of the proposals is done in parallel and then the results are compared). However, it is obligatory that in the evaluation there be peer review and the discussion of proposals in expert councils.

Russian Participation in Decision Making by Foreign Foundations and Organizations

To what degree does the Russian side participate in decision making about proposals submitted by Russian researchers to foreign foundations and organizations? In general there is the following rule: if the projects are financed only by foreign organizations, peer review is conducted by Western experts. If the program envisions co-financing, then regardless of the amount, the Russian side participates in making the decision about projects. However, there have been some exceptions to this general rule, such as the individual research and writing grants of the MacArthur Foundation, which were financed entirely by MacArthur but were evaluated by joint Russian-U.S. selection teams.

The Russian side is a rather active participant in the selection of projects and the administration of programs by foundations having offices in Russia. The presence of such offices in Russia means that for the given foundation or organization Russia is a priority, and not simply an object for support or cooperation along with other countries.

A second generalization is that the existence of co-financing from the Russian side always means Russian participation in selection of projects and/or their administration. Also, for organizations that distribute very large sums in Russia (ISTC, INTAS), the Russian side participates in the making of decisions regardless of whether or not there is co-financing. In a few cases when the co-financing is modest (10–15 percent of projects), the final decision is made by the foreign organization (for example, NWO).

There are no essential differences between U.S. and European programs on the question of inclusion of Russian participation.

Although tables 6.4–6.6 demonstrate considerable Russian involvement in the activities of foreign foundations and organizations on specific projects, the decision about what general programs to support in Russia and how to do it is made independently by 75 percent of the organizations and foundations studied, without consultation with the Russian side. The missions of foundations in Russia are formed either on the basis of the general orientation of the specific foundation or through its own analysis of the current situation in Russian science and scholarship.

Russian government agencies gain influence on the activities of foreign science programs in Russia primarily through co-financing of projects or other forms of support (for example, as in the case of the ISTC, buildings and real estate). Such participation obviously demonstrates Russian commitment to the initiatives and appreciation of their value.

The Evolution of the Goals and Priorities of Support

The goals of foreign organizations and foundations in Russia have changed; one can periodize the goals and priorities in the post-Soviet period in the following way:

Table 6.4. Participation of Russians in programs of international and U.S. foundations

	ISTC	INTAS	CRDF	NSF (USA)	MacArthur Foundation
Requirement for cooperation with foreign scholars	Not required, but welcomed.	Yes. There must be several participant countries taking part in each project.	Depends on program. In joint research programs and also in commercialization projects there must be U.S. partners.	Required. Russian scientists are not supported by themselves. They can receive small grants for equipment and receive pay as consultants.	No
Nationality of project leader	The leader of the project is a Russian scientist, but foreign colleagues may be partners or consultants.	Depends on the initiative. In scientific projects the leader cannot usually be from Russia, but in grants for young scientists the leader is a Russian scientist.	Depends on the program. In joint research programs and in commercialization projects there must be co-leaders from the Russian and U.S. sides. In other projects the leaders are Russian scientists.	The project leader must be a U.S. scientist.	Russian scientists or scholars.
Participation of the Russian side in the selection, administration, and/or financing of projects	The Russian side participates in the selection and administration of projects but not in financing.	The Russian side participates in selection, but not in administration of projects. Co-financing is possible, but the funds must be transferred to INTAS accounts.	Only in several selected programs (BRHE).	Russian scientists usually do not participate in any procedures except expert review.	The Russian side participates in the selection and administration of projects, and in programs of science and education centers also in co-financing projects.

Source: I. Dezhina, interviews, 2003–2007.

Table 6.5. Russian participation in the programs of individual European nations

	NWO (Netherlands)	British Council	Organizations and foundations of Great Britain (Wellcome Trust, Research Councils, Royal Society)	German Academic Exchange Service (DAAD)	German o-ganizations and foundations (von Humboldt, DFG, Max Planck Society)
Requirement for cooperation with foreign scientists	Yes	No	Yes, since most of the organizations give grants for trips to Great Britain for joint research.	Yes	Yes
Nationality of project leader	The project leader is a Dutch scientist, but on the Russian team there must be a leader and chief partner.	The project is always carried out in Russia by Russian scientists.	In certain cases a Russian scientist is the leader of the project (Wellcome Trust)	This concept is not applicable.	In cases where this concept is applicable, the leader is a German scientist and a Russian is a co-chair.
Russian participation in selection, administration and/or financing	Russians participate in all activities: selection of projects (through the RFBR), administration (RFBR), and co-financing (ministries or RFBR).	Russians participate in all activities.	Russian scientists participate as experts (Research Councils).	Russians participate in selection of proposals.	No, although the possibility of the participation of Russians in selection of projects and also in co-financing of individual initiatives is being discussed.

Source: I. Dezhina, interviews, 2003–2007.

Table 6.6. Russian participation in humanities and
social sciences programs financed by the U.S. government

	IREX	Fulbright
Requirement for cooperation with foreign scholars	No	No
Nationality of project leader	Russian scholar	Russian scholar
Russian participation in selecting, administering, and/or financing projects	Russian scholars participate in the second stage of project selection by interviewing the applicants. Administration is partially shared with the Russian government. However, all political and strategic decisions are made in the United States.	Russian scholars participate by reviewing projects and interviewing applicants.

Source: I. Dezhina, interviews, 2003–2007.

- 1992–1994: A period of emergency first aid to Russian science—the support of scientists and leading scientific groups, aid to libraries and preservation of valuable materials and collections.
- 1995–1997: Growth in cooperative research by Russian and foreign scholars, the appearance of the conception of parity and shared financing of R&D, the continuance of support for infrastructure (especially telecommunication projects). Growth in support of specific categories of grant recipients: young scientists, university scientists, researchers in regional areas.

In this period foreign foundations demonstrated a tendency toward a transition away from individual and group grants (what was often called "pure philanthropy") to cooperative projects with the participation of foreign partners. Although these projects were often mutually beneficial, the larger part of the financing (75–80 percent) was by the foreign organizations or partners.

Such cooperative projects were more selective and preferable because they presupposed not only a high level of quality but also the establishment of contacts with Western partners. As a result, a more limited number of select institutes participated in cooperative projects. Only the leading scientific centers could meet these higher criteria. Furthermore, cooperative projects were fewer in number than earlier philanthropic ones—on the average in each program fewer than 200 projects were financed in the course of two or three years. Since the support of the strongest scientific groups fostered stratification among scientists and institutes, the foundations more and more often began to create preferential conditions among "the strongest"—geographic distribution, youth, women.

- 1997–2002: A growth in initiatives aimed at reforming Russian science, including creating new infrastructure and stimulating connections between research and teaching and between research institutions and industrial enterprises that were potential users of R&D. In this period even more attention was given to support for young scientists, university researchers, and researchers outside Moscow and St. Petersburg.

In 1997–1998 a systematic reorientation of the activity of foundations began: the period of first aid was finished, and foreign organizations more frequently began to work for the reform of science in Russia. The basic efforts were the following: the development of the information infrastructure of Russian science (especially in telecommunications and internet access, libraries, subscriptions to journals, new textbooks), attempts to bring research and teaching closer together, and also efforts to bring theoretical and applied research into closer contact.

Another sign of the reduction of philanthropic support was the fact that more and more foundations began to insist on shared financing of projects (sometimes on the basis of parity) by Russian state and regional organizations. Thus in 1997 cooperative competitions were announced by RFBR and INTAS, RFH, and INTAS; the budget for the latter was about €9.9 million, including contributions of both sides. Also a program of the RFBR and the DFG (German Research Foundation) was announced in which each side paid the expenses on its own territory. Some of the already existing programs began to develop on a new financial basis, with the attraction of funds from local regions (the International Soros Science Education Program—ISSEP—and the program for regional internet centers of the Open Society Institute).

Foreign foundations played a special role in the reforming of science and education activities in Russia. At the center of attention was the problem of combining science and education, bringing research and teaching closer together. This problem became pressing after the disintegration of the USSR, when a search for a new model for the organization of science began. The place of scientific research in universities in Russia was shrinking at the same time that the universities, paradoxically, were playing a larger and larger role in preparing high quality specialists. In 1991 the universities prepared about 60–70 percent of the graduate and doctoral students; by 1997 the figure was higher than 80 percent.[21] (The rest were prepared by specialized institutes, such as those of the Academy of Sciences.) As a result, about a fourth of all graduate students in universities had no opportunity to do scientific research, along with a similar portion of professors and teachers.[22] In this period a Russian-U.S. program began for the creation of science and education centers in Russian universities (BRHE).

- 2003 to the present: This period brought a growth in the variety of initiatives, the development of programs directed toward a broader inclusion of Russia in various international projects, and an increase in the number of business-oriented programs for helping commercialization of Russian R&D. A majority of foundations continued their work and diversified their activity in Russia. At the same time there are a number of organizations—first of all, those giving grants for graduate study of Russians abroad (Fulbright Program, German Academic Exchange Service, Alexander von Humboldt Foundation)—that did not change the scale and direction of their work.

During this period a growing number of European organizations and foundations included Russia in general European programs in science and technology. As a result, there was a predictable increase in the average size of grants for cooperative projects at the same time that the overall number was decreasing. Another characteristic was greater concentration on specific directions of research and on special projects, including cooperative work on global problems. For European organizations the themes of the 6th Framework Program of the European Commission became priorities.[23] Europe, openly declaring its competition with the United States and Japan in the sphere of innovation, stated its wish to use Russian (and Ukrainian) potential in this global competition. Europeans were taking official steps in the highest priority areas of R&D, such as nanotechnology and new materials. However, Russia up to the present has not been able to be an equal participant in the projects of the European Union. First of all, Russia is neither a member of, nor an applicant to, the European Union. Second, in comparison with Western scientists Russian applicants have so far been inadequately competitive in their ability to write proposals that lay out a research and administration plan for a project, distribute obligations among researchers, and give a description of results they propose to achieve. It is no accident, therefore, that the level of success of Russian projects in the competitions of the 6th Framework program was so far not high, only about 12 percent.

U.S. and international organizations, including the ISTC, have been focusing more sharply on priority areas. Earlier they supported a greater variety of disconnected projects, but recently they have often introduced a "programmed approach" emphasizing certain themes, such as "alternative sources of energy." Under such themes a competition is announced for the fulfillment of concrete assignments, which is accomplished by several institutes or groups.

Contemporary Tendencies in Foreign Financing of Russian Science

During recent years several new tendencies have arisen in the activities of foreign organizations and foundations in Russia. The first tendency is the contraction of the absolute size of financing by foundations, accompanied by a growth in contracts between Russian research groups and foreign organizations. Thus the total share of Russian research financed by foreign sources has remained fairly stable, around 9.5 percent, but grants have diminished while contracts have increased.[24]

The share of foreign financing of Russian science may also be diminishing because of the growth of internal sources—first of all, industrial funds, although the business sector still supports R&D in Russia at a rather low level (7.5 percent in 2004),[25] and the share of industry in expenditures on science has not recently grown (for comparison, in Western countries the share of industry in supporting R&D varies from 50 to 75 percent). Among the former socialist countries Russia is close to Hungary in the share of foreign financing of its science (about 10 percent), and that share substantially exceeds all other such countries (Poland, 4.8 percent; the Czech Republic, 2.7 percent; Slovakia, 2.1 percent).[26] However, it should be noted that the share of foreign investment in Russian science is not as large as that of several other countries of the world; the share of foreign financing of science in Austria is 18.5 percent, and in the UK 20.5 percent. On the other hand, in the United States foreign sources for the financing of science are practically absent.

Starting in the early 2000s the funding of Russian science from the U.S. government in the Nunn-Lugar Program and the Freedom Support Act (FSA) has steadily decreased. Correspondingly, the budgets of foundations receiving their funds completely or partly from the government have also decreased.

This trend is connected with changes in the tasks the United States faces. From the beginning large sums were invested in Russia to convert military science, stem brain drain, construct a civil society, and support the social sciences. Presently it is recognized that in all these areas diminishing returns have set in. Furthermore, the United States now has more important tasks than the support of Russian science (for example, the struggle with terrorism). The regional interests of the U.S. government have also shifted (there is great interest at present in the support of science and education in countries like Ukraine, Georgia, Moldova). Russia is losing its attractiveness in comparison with the countries of the CIS, and within Russia itself foreign organizations give more and more attention to regions outside Moscow and St. Petersburg.[27]

The chief areas that U.S. government sources continue to support in Russia are conversion in various forms and scientific exchange programs. And the exchange programs are more and more oriented toward Presidential initiatives. At the present time the struggle with AIDS and with terrorism take priority. Therefore exchange programs are more and more based not on the scientific disciplines, but on the themes emphasized at the governmental level in the United States.

Moreover, administrators of exchange programs are planning to increase the number of participants at the expense of the length of time exchangees spend abroad.[28] This trend is a reaction to the changing preferences of scientists and potential grant recipients themselves: those who actively work in science usually do not want to leave their country for a long time, fearing missed opportunities at home. To a certain degree this is a reflection of globalization, which is accompanied by an acceleration of all processes, including the production and spread of scientific knowledge.

During the last several years there has been a reexamination by Western administrators of support programs for Russian science (what to support, how, and in what conditions). New emphases have emerged. One of the new emphases is greater coordination and more cooperation among all the actors: Western foundations, donors, scientists, businesspeople; Russian administrators, scientists, businesspeople. Philanthropy from the Western side has diminished in importance, and mutual benefit has increased, as well as hope for both philanthropy and investment from the Russian side. Western actors want agreement with Russian actors at the beginning of initiatives. More organizations want to involve Russian scientists and scholars as experts and consultants in their programs. Flowing out of this move toward greater cooperation is increasing reliance on shared financing by private businesses and local governments.

Since 2002 there has been a Forum of Donors in Russia. This is a Russian organization (a non-commercial partnership) with its own charter, mission, and strategy of development.[29] One of the tasks of the forum is helping the establishment of Russian philanthropy and helping and coordinating the activity of foreign organizations.

A consequence of the strengthening of interactions of the various actors is the appearance of programs that are co-financed by several organizations (for example, the cooperation of the MacArthur Foundation, the Carnegie Corporation, CRDF, and the Russian Ministry of Education and Science). But so far the number of such programs is small, since arranging cooperative projects with the participation of several foundations and organizations is not simple. The various donors often have different understandings of the current situation and priorities, control over use of funds is complex, and different donors often have different systems of accountability.

There is growing interest in areas where support can bring visible, practical results. Consequently, more organizations are giving attention to the commercialization of the results of R&D and the development of partnership ties between Russian science and Western small businesses. The CRDF has broadened its seminars on the commercialization of technologies. The ISTC is investing more funds in the study of patenting and commercialization of the results of R&D.

As before, interest remains high, both on the Russian and the Western sides, in supporting young scholars and graduate students. Interest in possible reforms in science also continues, directed often toward creating research universities, "centers of excellence," and innovation infrastructures. The change in status of the Academy of Sciences in December 2006 attracts attention to possible further reforms combining teaching and research.

International support has been a major factor influencing science in Russia, not only in helping it at the moment of greatest need—in the nineties—but also in providing models for new ways of funding science. The ISF, one of the earliest foreign organizations to be become active in assisting Russian science, was especially important in introducing the principles of peer review and competitive funding. In later years, organizations such as INTAS and ISTC continued to make large sums of money available to Russian scientists while also supporting peer review and competition.

As of the spring of 2007, special support by foreign foundations and organizations for Russian science was declining. The cessation of new projects by INTAS starting in January 2007 was one indicator of that decrease. The reasons for the diminishing of support were not uniform. Some foreign foundation officials believed that with the recovery of the Russian economy they should shift their priorities elsewhere. At the same time, the Russian government indicated that it preferred agreements that were "mutually beneficial" over those that were "philanthropic." However, only about 20 percent of the foundations we are studying either are preparing to end their work in Russia in a given year (usually from 2007 to 2010) or have actually already stopped their projects in S&T. Officials from these foundations say that within a few years the mission of the foundation or program will have been fulfilled. Others say that within several years Russia will become an equal partner in international programs and therefore the need for special initiatives for Russian science will no longer exist.

Other foundation officials call for continuing work in Russia. The most commonly cited reason is the benefits of cooperation, both scientific and commercial, that come from working with strong Russian research groups. From an economic point of view such cooperation is often advantageous for foreign partners because salaries in Russia are lower than in many other countries with strong scientific establishments. Conducting cooperative research and even financing it is often less expensive for foreign partners than if they do the work themselves. Interest in gaining access to certain forms of high technology is still strong, and is emphasized by a number of foreign organizations (ISTC, CRDF, NWO). And a new concern is cooperation against terrorism.

Another reason given for continuing such work is the geopolitical importance of cooperation with Russia. For the European Union Russia is a geographical neighbor, and for the United States an important potential strategic partner. Fostering mutual understanding between such countries is an important goal, and the support of the academic community has turned out to be an effective means of doing so.

In addition to these main reasons for cooperative S&T programs (mutual scientific, commercial, and geopolitical benefit) a number of the older reasons remain valid: supporting civil society in Russia, strengthening a market economy, stemming brain drain, improving environmental ecology.

7 Strengthening Research in Russian Universities: A U.S. and Russian Cooperative Effort

One of the most interesting reform attempts in Russian science in recent years has been the effort to strengthen research in Russian universities. To an American, accustomed to research universities, this attempt does not seem innovative or radical. But in Russia the combining of research and teaching in universities is a much more novel idea. Traditionally, Russian universities have been primarily teaching institutions; high-quality fundamental research was usually performed in the institutes of the Academy of Sciences.[1]

Creating a system of research universities in Russia of international rank would be both an enormous task and a positive reform. In discussing the desirability of such a reform in chapter 1 under World Trends in the Organization of Science, we noted that universities are increasingly seen throughout the world as especially productive loci for research. In the best of circumstances, teaching and research exist in a symbiotic relationship with each other.

The Basic Research and Higher Education (BRHE) Program

Probably the most influential effort directed toward uniting teaching and research in Russian universities has been one initiated from abroad: the Basic Research and Higher Education program (BRHE), begun in 1997 by the John D. and Catherine T. MacArthur Foundation and administered by the Civilian Research and Development Foundation. Both these foundations have headquarters in the United States.

The first version of the BRHE program appeared at the end of August 1997 as a result of an initiative of the MacArthur Foundation, which directed to a U.S.-Russian group of specialists the question, "If a philanthropic foundation wants to support fundamental science in Russia, what would be the best way to do this?" The committee concluded that the most innovative option was to support research in Russian universities in a way that would allow them to compete with the institutes of the Academy of Sciences.

Research and Education Centers (RECs)

The strategy chosen to accomplish this goal was to establish Research and Education Centers (RECs) within Russian universities. The RECs were asked to promote high-quality scientific research and education while cooperating closely with other leading research, education, government, and community institutions (including institutes of the Academy of Sciences). They were offered grants large enough to purchase state-of-the-art equipment for research and teaching that might serve as magnets to attract both established and young researchers. Furthermore, three components were to be combined in the RECs: education, research, and the development of external ties with scientific, educational, industrial, and other native and foreign organizations and enterprises. Young investigators were assigned a high priority; the RECs were required to spend at least 10 percent of the award on students (both undergraduate and graduate) and junior researchers.

Western experience played an important role in the conception of the BRHE. The best research universities in the United States served as models. Also, in the United States the National Science Foundation has financed somewhat analogous efforts in universities (Science and Technology Centers).[2] In the NSF-financed centers the same three mutually interdependent components are emphasized—research, education, and external ties.

The BRHE program was introduced in Russia, but not as a solely U.S. program. The Russian Ministry of Education (after 2004 the Russian Ministry of Education and Science) joined in the effort and became a co-administrator, along with the Civilian Research and Development Foundation of the United States, to which the MacArthur Foundation (and later the Carnegie Corporation of New York) assigned budgetary funds. BRHE became a fully cost-shared program, receiving approximately 50 percent of its funds from the U.S. side and 50 percent from the Russian side (the Ministry supplied 25 percent, and local university, government, and industry sources supplied another 25 percent). The goal was for the Russian side gradually to take on more and more of the financing until the point came when the U.S. organizations could cease funding. By 2006 this process had proceeded to the point that the Russians were financing 75 percent of the program; the U.S. entities then announced that within a few years they would pull out entirely, hoping that the program had become sustainable on its own.

A special characteristic of the BRHE program was the constant monitoring and analysis of the results at each stage, including not only annual reports by the grant recipients but also annual visits by external experts and staff members of the program. Such visits raised the quality of management and also helped the RECs to analyze their activities and introduce corrections where possible.

The RECs were chosen competitively on the basis of proposals submitted by Russian universities. The RECs are designed to assist the universities hosting them to become modern research universities—a very tall order in the Russian context.

At first the large cities of Moscow and St. Petersburg were excluded from the competition as an effort to help provincial universities, but later the competition was opened to all universities in Russia. By 2005 BRHE had established sixteen RECs across Russia: one in Moscow, two in St. Petersburg, and thirteen in other regions of Russia. The sixteen winners are listed below:

Nizhnyi Novgorod State University
Krasnoiarsk State University (after 2006, Siberian Federal University)
Far Eastern State University (Vladivostok)
Rostov State University / Taganrog State University of Radioengineering
   (joint award) (after 2006, Southern Federal University)
Ural State University and Ural State Technical University (Ekaterinburg)
Saratov State University
Kazan State University
Novosibirsk State University
Perm State University
Voronezh State University
Moscow State Engineering Physics Institute, Moscow Institute of Physics
   and Technology (joint award)
St. Petersburg State University
Petrozavodsk State University
Samara State Aerospace University
St. Petersburg State Mining University
Tomsk State University

In total in the period 1999–2005 about $45 million U.S. was distributed for the support of these Centers, of which 51.3 percent came from U.S. sources, 25.8 percent from the Russian central government, and the remaining 22.9 percent from local sources. In 2005 the MacArthur Foundation designated an additional $10 million for the continuation of the program to 2010.

In the fall of 2006 four new RECs were selected, making a total of twenty. The new ones were:

Irkutsk State University
Baumann Moscow State Technical University
Tambov State Technical University
Demidov Yaroslavl State University

Core funding for the four new RECs comes entirely from the Russian side (the Ministry of Education and Science and local Russian sources). The U.S. side may later supplement the Russian grants to these four universities with small ones for certain activities, but the major funding responsibility is on the Russian side.

The fact that almost one-fourth of the funding for the sixteen RECs established by 2005 came from local sources in or near the university cities in Russia had great significance from the point of view of the philosophy and long-term goals of the program. The willingness of local sources to provide financing— regardless of whether these sources were the universities themselves, local governments, industry, or other benefactors—illustrated that the centers were needed in the given region and, correspondingly, had good prospects for sustainability after the termination of external financing.

Annual meetings (called "Pan-REC Conferences") allow the RECs to learn from each other's experiences. At these conferences students present exhibits of the research they have carried out under the BRHE program, and administrators discuss mutual problems. Other activities include summer English language immersion camps for students (in recognition of the fact that English is the international language of science today), a post-doctoral research and teaching fellowship program, training in research management, mini-grants, and competitive equipment grants.

In 2003 four competitive grants to create Technology Transfer Offices (TTOs) were established in host universities that have RECs. In 2006 four more TTOs were created. TTOs have been discussed in a previous chapter. In addition, training in technology transfer has been made available to all RECs.

Some Characteristics of the Existing Research and Education Centers (RECs)

There is no one model that all Centers try to follow, even though there are certain rules all Centers must observe. It was thought that creating a single template might suppress original initiatives.

Differences among the Centers are connected with the themes of research conducted in them, which often dictate the optimal size of research groups, the number of supporting personnel, and the kinds of necessary equipment. Research areas differ among themselves in the rates at which they produce publications and also in the frequency of conferences held to discuss results.

The number of scientific and teaching personnel in the Centers varies from 30 to 100 people. The variation in the number of students—undergraduate and graduate—is also great, from 25 to 150 people. Particular emphasis is placed on the involvement of undergraduates in research, since it is this characteristic that most distinguishes RECs from institutes of the Academy of Sciences (which have graduate students but usually few or no undergraduates). Some of the students receive financial support from the Centers while others voluntarily participate in Center activities on a non-pay basis. The maximum number of undergraduate and graduate students receiving financial support does not exceed 70 people per Center.

One of the favorable characteristics of the Centers is the proportion of young researchers; the share of scientists below the age of thirty-five (24 percent) is larger than the share of older researchers and teachers (17 percent). Furthermore, the ratio of teaching and research personnel to students, undergraduate and graduate, is approximately 1:1. Such a ratio permits a more individualistic approach to all students participating in the Centers and helps in the preparation of high-quality specialists.

Most of the RECs have improved during the life of the program. The administrators of the Centers have acquired clearer conceptions of their assignments and goals. Management has improved, as have the rates of publication in internationally peer-reviewed journals, a good indicator of quality.

How do the Centers spend their grant money? Average indicators for 1999–2004 show that about half of the total grant was spent on purchase of equipment.[3] Other expenditures were salaries (20 percent), support of youth (10 percent), travel expenditures (8 percent), and institutional overhead (10 percent).

The heavy expenditures on equipment (more than 2½ times those on salaries) requires an explanation. Universities in Russia have traditionally been disadvantaged in equipment relative to institutes of the Academy of Sciences. The budgets of universities, distributed by the government (through the Ministry of Education and Science), have been based on the number of students enrolled in the particular university, not on research needs. The budget of the Academy of Sciences, on the other hand, while also coming from the government as a total sum, has been distributed by the Academy administrators on the basis of perceived research needs, and therefore equipment has had a higher priority in the Academy than in the universities. Recognizing this disparity, the BRHE program has put heavy emphasis on equipment in universities with RECs. The first REC, for example (that at Nizhnyi Novgorod University), used much of its BRHE money to purchase and maintain a very expensive scanning probe microscope (an elaborate electronic installation occupying an entire room). Having such a piece of equipment meant that even nearby institutes of the Academy were interested in the university. For the first time, the relative positions of the university and the institutes with regard to equipment were the reverse of what was normally the situation. Building research universities of high quality in Russia can be enabled by such relationships.

Furthermore, such new equipment can be a connecting link between research and teaching, since students have access to and learn from research instruments that are on a level with international standards. In the past, university students were usually given access only to outdated teaching labs. Modern equipment, on the other hand, permits researchers and students to pose new scientific tasks and produce new scientific results that are then used for the updating of lecture courses. Such equipment also helps in establishing partnership relationships with other organizations (Academy institutes, industries, foreign research labs) and in conducting cooperative projects.

With the advantage of such modern equipment most of the Centers were able to arrange cooperation with Russian and foreign colleagues. In that way the Centers attracted additional financing for carrying out research, organizing conferences, and other undertakings. The Centers attracted, beyond the initial grant financing, on average about $200,000 a year. The additional funds amounted to almost 40 percent of the basic grant. In almost half of the Centers the amount of additional financing attracted to the center was higher than the total amount of money given by the U.S. side (through CRDF).

Beginning in 2003 the BRHE program became more diversified. In addition to the basic financing there appeared subprograms, such as post-doctoral stipends for young scientists with the "kandidat" degree (one hundred such stipends were awarded each year, each for three years), a subprogram for creating Technology Transfer Offices; grants for small equipment purchases; and minigrants for other efforts (organization of conferences, improving the quality of university managers). All these initiatives were organized on a competitive basis. Each of the Centers received differing amounts of such supplementary support. Their amounts in 2003–2004 varied from $70,000 to $425,000.

Technology Transfer Offices (TTOs) in the BRHE Program

In December 2002 the CRDF announced a new initiative to open Technology Transfer Offices (TTOs) in those Russian universities where, as a part of the BRHE program, Research and Education Centers in the natural sciences have already been created. This step recognized that although at first the BRHE program focused on support of fundamental research, later it became necessary to take the next step and help universities in learning the skill of commercializing R&D results. This BRHE effort to establish TTOs in Russian universities relied heavily on foreign models, particularly the United States in the wake of the Bayh-Dole Act (1980).

The basic functions of TTOs were discussed in chapter 5. The benefit of a TTO for a university, based on foreign experience, lies not in the receipt of large royalties but in the preparation of specialists for the business sector and in developing ties between industries and universities that, in time, lead to grants and new R&D contracts.

The CRDF and the Russian Ministry of Education and Science conducted a competition after 2002 with the goal of creating in Russian universities TTOs similar to those in U.S. universities. Four winners were chosen (Nizhnyi Novgorod State University, St. Petersburg State University, Ural State University, and Tomsk State University). These four universities were given financing from the Ministry and CRDF at a ratio of 1:2 ($75,000 from the Ministry and $150,000 from CRDF for a term of three years).

The goals of the university TTO program are improving the quality of university management of innovation, changing the mentality of teachers and students about technology transfer, and strengthening economic ties between the university and the region in which it is located. The merits of the model are the close integration of the TTO with the university, rapid and simple access to research results, and also the possibility of reinvestment in additional scientific research.

International experience shows that TTOs inside universities are sometimes inadequately receptive to market signals. In Russia this problem is even greater than elsewhere. In Russia, many basic attitudes, policies, and laws related to a market economy are still unclear or inadequately developed, such as those concerning intellectual property, market research, industrial interest in university research, and engagement of private investors in innovation. Nonetheless, university administrators in Russia are often very interested in TTOs, perhaps seeing in them more benefits than can be realistically expected.

The TTOs in the BRHE program have been in operation only for about three or four years, and therefore substantial progress in commercializing R&D results cannot be expected from them. So far the study of markets has not been sufficiently intensive and there is as yet no income from royalties.

Most of the budgets of the four university TTOs are at present coming from the government (the Ministry of Education and Science), the "mother" university (often in terms of office space), and grants (including those from CRDF). Commercial activity is a small part of the work of the university TTOs; instead, most of their activity has been educational, legal, and organizational. However, the four TTOs have established more than four hundred contracts with industrial enterprises, and somewhere between 5 and 15 percent of their budgets come from these contracts.

Although the university TTOs have not yet produced much income, they have succeeded in creating a number of beginning innovative firms, varying from six to twenty-seven in each of the four. Whether these infant enterprises will continue to exist is, of course, unknown. There are few "incubators" or Technology Parks where they can find favorable conditions in their first stages of development, and their budgets are constrained.

In the next section we will evaluate the BRHE program in general, not the TTOs, which are still too recent for evaluation.

Results and Effects of the BRHE Program

The BRHE program has been an unusually effective reform effort in Russian science, one successfully combining both foreign and local inputs.[4] It has demonstrated that the idea of creating "research universities" in Russia is not unrealistic, even though the effects so far are much too small and limited to fulfill that lofty goal. The effort is best seen as a successful pilot program that may or may not be followed by a more ambitious effort at replication and expansion. To fulfill the goal of creating a system of research universities in Russia would require both financial inputs to universities and policy changes concerning them that so far the Russian government has not been able or willing to execute.

The purchase of expensive research equipment as a centerpiece of the BRHE program permitted not only the development of new lines of research but also a change in the relationship of the Academy of Sciences and the universities. Many researchers from institutes of the Academy participated in the Centers, and in some Centers workers from the Academy were the directors of research projects. Cooperation became more equal, and interest in cooperative work was increasingly shown not only by university researchers but also by those in Academy institutes. In that way an effect was achieved that was not attained with the Integration program to be discussed later in this chapter.

The BRHE program made an important financial input to Russian universities at a time when government financing was both minimal and irregular, and it did this in a way that gave stability, guaranteeing financial support for five years. This timely support had concrete results: on average each REC produced about 100 scientific articles every year, of which about a third were co-authored with young researchers in the Centers. They also prepared about 130 reports at conferences, of which 40 percent were co-authored with students. On average each of the participants in the Centers had 0.5–2 articles in refereed journals, and the share of works published in international refereed journals was growing.

As a result of the synergetic interaction of three components—education, research, and the development of external ties—the image of the Centers within each university was improved, and, eventually, the entire university benefited. The sixteen universities where RECs are located acquired an elite status. Many other universities now wish to have this status, and there is a danger that the title Research and Education Center will be devalued by imitation and replication unaccompanied with sufficient financial input or quality control.

Within the sixteen RECs new research topics appeared, and interdisciplinarity was strengthened as a result of the modernization of equipment and the improvement of the material situation for research. These improvements led to more cooperation of the universities with each other (facilitated by the Pan-REC conferences) and with other scientific organizations, and the interaction of different faculties within universities.

The BRHE program also affected the curricula of the sixteen universities, simulating the updating and reexamination of new courses and laboratory instruction, and drawing in new students. All the Centers developed new textbooks for courses and even opened up new specializations based on the scientific results obtained with the recently acquired equipment. Scientists from other institutes and regions were often invited to give lectures to undergraduates and graduates, and also for conducting cooperative work. A number of Centers conducted new recruitment into master and doctoral programs in which not only students and graduates of the given university could enroll but also anybody wishing to do so from other regions.

All the above effects were positive, but it must be admitted that the lasting impact of the BRHE program is still unknown, as will be discussed in the next section. With the cessation of foreign support and the effort by other universities to imitate the model created by the RECs, it is entirely possible that traditional managerial attitudes will take over and that the influence of the reform will become less and less visible. Also, it must be recognized that the BRHE program has not stopped brain drain from the institutions in which RECs are located, and may actually have slightly stimulated it. In the strongest Centers students who early began to do research, to publish, to learn English, and to go abroad to present their results at conferences were able in that way to evaluate their ability to compete with foreign colleagues; as a result, the best of them often began to emigrate abroad more energetically. However, the relative scale of this outflow was not great, and there are some ways in which it could be abated. The emphasis of the RECs on equipment, not on salaries, while a valuable step in terms of the time, meant that talented researchers had opportunity to improve their salaries markedly by going abroad. If salaries at their home institutions could be improved, the temptation would be less.

Change in the BRHE Program

Beginning in 2005 the BRHE program was significantly modified, both in basic principles and in the amounts of funds being distributed. Instead of "base" support of the Centers a transition was made to project financing. The administrators of the program, directed by a joint U.S.-Russian committee, planned to conduct four competitions in 2005–2008, and on the basis of the results there would be given out thirty-two grants, each of $200,000. The funds can be spent both on scientific and educational activities, and, as before, 10 percent of the funds must support young scientists and undergraduate and graduate students. In that way a policy of supporting the strongest was chosen, an approach that could help the process of deeper reform of science.

A second essential change, as earlier mentioned, concerned the sources of financing of the Centers. Earlier the share of the U.S. funds had been 50 percent; now it was 30 percent. At the same time the share of the Russian Ministry of Education and Science rose to 35 percent, as did that of local sources. Under this scheme, therefore, 70 percent of all money came from Russia, not the United States. The U.S. sponsoring foundations also notified the Russian participants that in 2010, independent of the results achieved, they would cease financing the program. An effort to attain sustainability of the Centers on the basis of funds only from Russia was clearly being made.

This change in the program caused strong concern in the Centers, since quite a few of them doubted that they could attract sufficient funds to continue operation at the current level. Moreover, some recent changes in legislation seem to make co-financing by federal and regional sources of scientific and innovation activities significantly more complicated than earlier.

Prospects for the Centers after BRHE Program Financing Ends

How sustainable are the Centers of the BRHE program?

A survey was made in May–June 2005 of the sixteen Centers, which revealed the deep concerns of the leadership of the Centers about their future sustainability. The survey, conducted by one of the authors, showed that the leaders of a large majority of the Centers (fifteen out of sixteen) believed that their Centers would be sustainable only under certain conditions that were by no means guaranteed. Those conditions included not only improved support of science in Russian universities by the central government and local authorities, but also a continued co-financing by the BRHE program itself, an apparent contradiction to the U.S. foundations' announcement that they would cease financing in 2010. However, the leadership of the Centers apparently hope that new Russian foundations will step in, a hope for which so far there is little evidence.

The leadership of two centers believed that sustainability could be achieved by integrating universities and organizations of the Academy of Sciences. Such integration has been much discussed, and the changes in the status of the Academy in late 2006 (discussed in chapter 3) possibly could facilitate such integration, but, again, so far there is little concrete evidence that this will happen in a way that will substantially help the existing state universities, including the sixteen with RECs.

In the absence of such rather dramatic changes, the main hope of the Centers for sustaining at least part of their activities is in reorienting themselves away from basic research toward applied and commercially significant research. Whether they could accomplish this in a sustainable way is by no means certain, and even if they did it would mean turning away from the original goals of the program as graphically described in its title: *Basic* Research and Higher Education.

As earlier mentioned, quite a few Russian universities that never had a REC (some applied in the competitions but were turned down) would now like to create one. However, the leadership of the existing RECs is skeptical about such duplication elsewhere. Only about 30 percent favor such efforts. Half of the Centers think that duplication is in general impossible, and the rest believe that only partial duplication would be possible. The reasons they give for these difficulties is the belief that each Center is unique in its structure, approaches to solving problems, and level of qualification of personnel.

Approximately half of the Centers believe that partial duplication would be possible. What they mean by "partial duplication" is the adoption of certain elements of the BRHE model, such as the grant system of encouraging young scientists, emphasis on educational programs connected with research, centers for the shared use of equipment, and analogous centers in the social sciences (a topic to be discussed in the following section). Such partial duplication may be the most realistic hope.

The single largest obstacle to sustainability and duplication of the Centers, all the leaderships agree, is financial. The BRHE program provided funds which, at the moment at least, do not seem to be available without it. The Ministry of Education and Science, the university administrations, and the local sources of funding (all of which provided partial funding for the Centers in the past) will thus be faced with a basic challenge. In the improving economic conditions of Russia, will they be able to provide full funding for what almost everyone agrees was one of the most successful reform efforts in Russian science and education? Unfortunately, the virtue of the effort seems to be no guarantee of its success.

The Centers for Advanced Study and Education (CASE) Program

In 2000 the Carnegie Corporation of New York initiated a program in Russia to promote interdisciplinary research in the social sciences and humanities. Although in this book we are concentrating on the natural sciences, the CASE program in the social sciences and humanities deserves attention because it coincided with the goals of the BRHE program in building up the research potential of Russian universities. In fact, there was an overlap in that six universities eventually had both CASE and BRHE centers and therefore were strong candidates for the status of research universities.

The CASE program was preceded by a detailed analysis of research in the humanities and social sciences in Russia carried out by the Kennan Institute for Advanced Russian Studies of the Smithsonian Institution on the basis of a grant from the Carnegie Corporation of New York. Organization of the program continued until 1999, and then the initiative was joined by the MacArthur Foundation. The program actually began in 2000. It has come to be known as the program for Centers for Advanced Study and Education, or CASE.

The goals of the program are the broadening and improvement of research in the social sciences and humanities in Russian universities, and the establishment of closer ties between Russian scholars and their foreign colleagues in these fields. These goals were determined by the Carnegie Corporation of New York in partnership with Russian government agencies (especially the Ministry of Education and Science).

Nine interdisciplinary institutes of the social sciences are a central element of the program, located in the following government universities:

Voronezh State University
Far Eastern State University (Vladivostok)
Irkutsk State University
Kaliningrad State University
Novgorod State University
Saratov State University
Tomsk State University
Ural State University
Rostov State University

The program includes both base financing of the CASEs and also grant competitions, including ones for young scholars. At first basic funds were directed at forming academic libraries. Then a grant competition for carrying out research in the CASEs was added.

At first all CASEs received the same base support. As the program proceeded it became clear that the institutes that had been created were very different, and therefore a decision was made to begin a gradual transition to "project" financing—that is, to the disbursement of funds on a competitive basis for different projects.

As the program has evolved, the educational component has become stronger. In the beginning the program leaders feared that if the CASEs concentrated primarily on educational activity, they would be converted into service centers of the corresponding departments and faculties of the universities.

The activities of the CASE program in Russia are administered by a philanthropic organization in Mosow, the INO-Center (INO stands for Informatsiia-

Nauka-Obrazovanie, or Information-Science-Education). The annual budget of the program is about $3 million, and of this about 25 percent is co-financed by the Russian Ministry of Education and Science. The U.S. funds are administered by the Kennan Institute for Advanced Russian Studies in Washington, D.C. The funds go to universities, including about $200,000 a year for the support of the CASEs. Support for the program was promised for four more years.

Not long ago the John D. and Catherine T. MacArthur Foundation joined up with the Carnegie Corporation of New York, the principal U.S. funder, with an additional grant of $1.6 million for the CASE program.[5] Thus, the MacArthur Foundation and the Carnegie Corporation have made an unusual alliance in the support of education and research in Russia; for the BRHE program, the MacArthur Foundation is the principal funder, and the Carnegie Corporation is the support funder; for the CASE program, the situation is the reverse.

The statistics of the CASE program demonstrate that grant funds are spent on the support of libraries (23 percent), conferences and seminars (21 percent), international projects including international travel (21 percent), publishing costs (15 percent), and institutional overhead (20 percent).

Each interdisciplinary institute has a curator—a well-known scholar in the area of specialization of the given CASE—who serves as a consultant to the staff of the CASE. The need for such curators arose because the social sciences, in contrast to the natural and technical sciences, are inadequately developed in regional universities. Here we see a difference between the CASE program and the BRHE program. For the CASE program it was necessary to attract scholars from Moscow and St. Petersburg to help in orienting research themes (from the point of view of topicality, originality, etc.) and improving libraries. In this way CASE has turned out to a significant degree to be a study or training project for the regions. It turns out to be much easier to help the natural sciences in Russian universities than to help the social sciences and humanities, a reflection of the fact that the latter fields were much more affected by ideology in the Soviet period.

In the future the CASEs should become centers that carry out multi-factorial research with the involvement of investigators of different backgrounds, and also a series of researches, seminars, written works, field research, presentation of results, inclusion of international partners, writing of collective monographs with foreign participation, and, finally, the conversion of new knowledge into study materials and texts.

BRHE and CASE—A Foundation for Forming Research Universities in Russia

Since the Research and Education Centers in Russian universities of the BRHE program were formed earlier, the CASE program has been seen by many analysts as a way of spreading the BRHE model. This view is strengthened also by the fact the U.S. sponsors of the CASE program were the same foreign foundations—the Carnegie Corporation of New York and the John D. and Catherine T. MacArthur Foundation. Both initiatives were taken cooperatively with the former Ministry of Education, and were continued by its successor, the Ministry of Education and Science.

Despite the fact that the situations in the natural and social sciences were different, the programs have quite a few common characteristics. At the beginning stages of the development of the Centers, support was give on the "base" principle, that is, all Centers received equal levels of financing. With time, separate initiatives began to appear (grants for young scholars, grants for purchasing small equipment, etc.) in which the Centers began to participate on a competitive basis. The development of this "programmatic" approach permitted the concentration of resources on the transformation of science and education in Russia, and therefore gave the Centers a special significance in the reform process.

Furthermore, in both programs a key component was the stimulation of ties between research and teaching. Such links were from the beginning one of the basic programmatic goals of BRHE, and were even more supported in CASE as it evolved. Both programs supported youth, rewarded mobility and an interdisciplinary approach to research, financed new equipment, and improved reference sources for research. Finally, the Centers developed connections with other organizations, both Russian and foreign, participated in cooperative science-education projects, and invited well-known scholars to give lectures—in a word, the Centers in their present form possess many of the characteristics of research universities. This is especially true in the six universities that possessed both CASEs and RECs, that is, have received grants from both programs. These universities are:

Voronezh State University
Far Eastern State University (Vladivostok)
Saratov State University
Tomsk State University
Ural State University
Rostov State University

The experience accumulated by the Centers may be more broadly extended to the universities where they are located. The chief sponsors of BRHE and CASE intend to continue support of the programs for the following three to four years (for BRHE, until 2010). Therefore, both for BRHE and for CASE there may still be time for participation in the formation of Russian research universities, or at least the beginning stages of such formation. Such an ambitious broadening of the programs can be effective because they have already successfully passed the "pilot" stage.

Still another current direction where the Centers can play helpful roles is connected with the signing by Russia of the Bologna Declaration aimed at creating a more uniform system of higher education throughout Europe (e.g., comparable degrees).[6] Some people in the Centers are not only familiar with Western universities and their organization, but also knowledgeable about the Bologna Process itself. Therefore the Centers can help transform the Russian system of higher education, explaining the ideas of the Bologna Process and also defining the mechanisms for their optimal use in Russia.

Universities where Centers have been created actively cooperate with organizations from the Academy of Sciences in their regions, attracting from them researchers to do cooperative scientific work and teaching. The Centers began to develop cooperation also with each other, especially those with intersecting research themes. And at the same time, beyond scientific cooperation, there is an effective exchange of experience in the organization of the Centers. In this way the Program differs essentially from its analog (to be discussed in the next section) named the Integration Program, which was created by the Ministry of Education of Russia, the Ministry of Industry and Science, and the Academy of Sciences.

The Integration Program

At the same time that the BRHE and CASE programs were attempting to bring teaching and education closer together in Russian universities, the Ministry of Education, the Ministry of Industry and Science, and the Academy of Sciences were jointly sponsoring their own program that seemed, at least by title, to be similar. This latter program began in 1996 under the title Integration of Science and Higher Education in Russia and was usually called, for short, the Integration Program. The tasks this program was supposed to accomplish were the integration of the research and pedagogical processes, the development of new specialties (including interdisciplinary areas), and the identification and support of the most able students. The first and most important task, the integration of the research and pedagogical processes, seems to be nearly identical with the aim of the BRHE and CASE programs. They too wanted to bring teaching and research closer together. On closer examination, however, we can see that the conceptions about how to do this were different. The BRHE and CASE programs were based on the principle that this integration should occur within the universities themselves. The Integration Program was based on the principle that the universities and the institutes of the Academy should develop links allowing closer interchange between the two systems, but that the basic function of the universities was to teach, while that of the institutes of the Academy was to do research. This is an essential difference from the BRHE and CASE programs, in which teaching and research are combined in the same institution, namely the university.

Another characteristic distinguishing the Integration Program from the BRHE and CASE was the scale of operation. The Integration Program was very large and ambitious, and included 154 Education and Science Centers facilitating interactions between the university and Academy systems. Within each of these centers were ten to fifteen different organizations. In other words, the

Integration Program was enormous, although supported with relatively modest sums of money. The Integration Program was stopped in 2005 when the government decided to put its priorities elsewhere, particularly into programs more similar to the BRHE.[7] Thus, the BRHE program demonstrated its strength relative to the Integration idea, a remarkable fact for a suggestion originating in the United States.

In order to evaluate the Integration Program, one of the authors in 2000–2001 conducted interviews in twenty-five of the Education and Science Centers of the Program. These Centers were located in Moscow, St. Petersburg, Saratov, Nizhnyi Novgorod, Vladivostok, Tomsk, and Krasnoiarsk. These cities contained Centers in the Integration Program that were thought to be among the most active and effective.

A number of interesting findings emerged from the interviews, many of them positive. The Centers were, indeed, promoting cooperation between universities and research institutes. Furthermore, students were encouraged to develop independent research projects at early stages. The preparation at the undergraduate and graduate levels was strong. Good preparation was offered in foreign languages, especially English. Students in the universities were often allowed to use the best equipment in the Academy institutes. All this special attention had the result that a greater number of university graduates decide to choose scientific careers, going on to graduate work in Academy institutes.

At the same time, a number of problems became apparent. First of all, the universities and the institutes of the Russian Academy of Sciences fall under different bureaucracies, and cooperation between them was often difficult. The universities are under the Ministry of Education and Science, while the Academy is directly under the central government. Funds were transferred to the chief organization in the project, often an Academy institute, and this organization often spent the lion's share on its own needs. Sometimes the funds were shared proportionately among the participants and then each organization spent them in its own way. If this happened, a plan connecting all the components of activity of the Center was not formulated, since there was no single administration. In several Centers councils were formed to give recommendations for cooperation, but without authority to direct the work. Therefore, there was often a disconnectedness between the universities and the institutes even though the declared goal was cooperation.

Researchers in the Academy institutes were sometimes willing to work with university students, but primarily in order to identify the very best students there in order to bring them over to the institutes, first as graduate students and later as junior researchers. The Academy seemed to define integration as the formation of "feeder channels" whereby the Academy could skim off the best students for its own research organizations.

Another major problem was financial. With so many Centers (154), the average level of financing under the Integration Program was usually so small (there were four or five favored exceptions) that the support was mainly of the moral variety rather than substantial. Modernization of equipment was often not possible except for minor repairs. A better approach would have been either to substantially increase the total budget for the Integration Program, or, failing that, to decrease the number of participating Centers. The Integration Program spread too little money too thinly over too many institutions.

Another financial concern was that grant distribution by regions was often seen as not being fair. Respondents outside the major cities of Moscow and St. Petersburg often complained that universities and institutes in those cities were unjustly preferred in the distribution of money. This complaint is controversial, since quite a few scientists maintain that the best Centers, in terms of quality, are located in Moscow and St. Petersburg and that is only natural that institutions there receive the lion's share of the funds.

Yet another problem was the danger of early specialization. In those Centers where a research institute, rather than a university, played the leading role, university students who came over to the institute to do research were often encouraged to specialize in a narrow area, the area most emphasized in that particular laboratory. Critics, often located in the university in question, maintained that such early specialization produced students with excessively narrow educations and interests and not suitably equipped for a life in which research emphases may change dramatically.

Despite these deficiencies, the Integration Program has had its successes. University students who took advantage of the opportunities offered by it to work in Academy institutes often improved and deepened their research skills and knowledge.

In the final analysis an evaluation of the Integration Program is affected by one's opinion about what the future organization of research and education in the sciences in Russia should look like. If one's goal is a system of research universities of the type found in the United States and some countries of Western Europe, then the Integration Program seems inadequate. It tried to preserve the essential characteristics of the existing university and Academy systems. If, on the other hand, one's goal is the improvement of the existing Russian system in which the universities and Academy have their separate functions, then the Integration Program, and others like it, have their value. It definitely creates closer links between the universities and the Academy in a way that often, but not always, benefits both.

As already noted, the Integration Program in 2005 became a part of the Federal Goal-Oriented Science-Technology Program, which assigns priorities in science and technology. Technically the Integration Program did not end at this time, but practically speaking it was no longer visible and no longer important.

The BRHE program has been, at least up to 2007, remarkably successful. Its success is most clearly reflected in its history: initially born at the suggestion of U.S. foundations and their advisors, it rather quickly won the support of the Russian Ministry of Education and of regional Russian authorities, who in the first years supplied 50 percent of all funds. In later years the Russian side began to supply an even higher proportion of funds, and in 2006 the Russian Ministry of Education and Science suggested a further expansion of the program; the ministry added that if the U.S. side could not provide more money, the Russian side would provide all funds required by the expansion. However, the Russian side also requested that the U.S. side continue to participate in the selection of the new Centers, even if it did not financially contribute to them. This joint cooperation in the BRHE program, both intellectual and financial, is a positive result of international collaboration in science and education.

Some of the reasons for that success are that the original idea, as advanced by the John D. and Catherine T. MacArthur Foundation, combined the components of research, education, external linkages, and emphasis on youth. The BRHE program has proven that Russian universities can effectively collaborate with Academy institutes; that integration of research and education is beneficial for both; and that the involvement of youth at an early stage in both teaching and research boosts their research careers in Russia and diminishes brain drain.

8   Impact of International Activities on Russian Science

As the two previous chapters have described, foreign foundations have pursued a variety of activities in Russia in science and education. This variety, however, can be classified in analytical categories. We identify five basic functions of the foreign foundations active in supporting Russian science: philanthropy, partnership, commercialization, security, and reform. Of course, not all foundations express their goals in these terms; rather, these functions are analytical categories created by us into which we have classified their work. Furthermore, not all the foundations have pursued all these functions. Nonetheless, we think that collectively their activities can be described in these terms. Let us look briefly at what goals the foundations have pursued under each of these five categories.

Under philanthropy, some foundations have aimed at improving the material situation of scientists, research groups, institutes, and universities in the former Soviet Union. They have helped upgrade scientific equipment and they have supported special categories of grant recipients: young scientists, undergraduates, and graduate students; women researchers; and scientists outside the capital cities ("regional science").

Under partnership, foundations have tried to support research enriching both Russian science and that of a partner country, usually the country of the donor.

Under commercialization, foundations have sought to use R&D in the interests of the private sector. Results include the appearance of new products and technologies, or at least the successful preparation of R&D results for commercialization (patents, licenses, contracts with private companies, etc.).

Under security, foundations have attempted to reorient scientists away from military research to civilian projects; they have also tried to prevent brain drain of scientists with knowledge of weapons to countries unfriendly to the donor country.

Under reform, foundations have helped the creation of new organizational forms of R&D activity, funded non-traditional centers of research, helped bring research and teaching closer together, and supported the formation of innovation infrastructures.

Of course the division of the functions of the foundations into the above five categories is only relative, since one and the same program can fulfill two and even more functions. For example, in the course of carrying out cooperative projects (fulfilling the partnership function) the material situation of Russian science is improved (fulfilling the philanthropy function). Obvious examples are the programs of cooperative research of INTAS, NWO, and CRDF.

"Indirect" results of the activities of the foreign foundations are the development of new management practices (for example, grant management), the spread of the system of peer review, and other administrative practices connected with the work of the foundations. The foundations made a definite contribution to the strengthening of the scientific community in Russia and also enriched the thematic structure of scientific work. Finally, foreign foundations have indirectly helped the appearance and development of native philanthropy in Russia.

Results of Philanthropic Support

The Improvement of the Financial Income of Scientists and Scientific Groups

The material situation of Russian scientists genuinely improved in a perceptible way during the time the foreign organizations were active, and at least partially because of them. For many Russian scientists foreign grants are more important than their official salaries because they are usually several times larger. At the present time the average salary of scientific researchers (including grants from Russian foundations) is about $300 U.S. a month (table 8.1) at a time when grants from foreign foundations usually give $400 or more a month. Moreover, foreign grants are not subject to income tax or value-added tax, and as a result the attractiveness of such grants is increased. (Table 8.1 shows not only the variation in pay of researchers in recent years but also the disastrous effect of the collapse of the ruble in 1998.)

The contribution of foundations to the support of former defense scientists was especially weighty: 52,000 such scientists received substantial additions to their salaries. Scientists received on average $25–30 a day, and this sum was not subject to any taxes. Therefore the participants in ISTC projects were able to restrict themselves to civilian research and usually did not seek other sources of financing.

At the present time the significance of the programs of foreign foundations for Russian scientists remains high. According to the evaluations of sociologists, at present a number of scientific institutes rely on grants from foreign foundations for a significant portion of their budgets—from 15 to 50 percent.[1]

Support of Young Scientists

In the beginning, organizations and foundations supporting science in Russia did not give special attention to young scientists and graduate students. Then they began to give preference to proposals including young people among the researchers. The next stage was the appearance of special programs for

Table 8.1. Average salaries of specialists carrying out R&D

	1997	1998	2000	2002	2004
Average pay from all sources in U.S. dollars	149	50	96	173	300
In percentage of average pay for all occupations in Russia	93.6	98.5	121.9	126.4	128.9

Source: *Russian Science in Statistics—2004 Statistichesky sbornik.* TsISN, Moscow, 2004, p. 86.

young scientists and graduate students (research grants, travel grants, stipends for study) for which youth received special support.

In the ISF long-term grants program, which was not, in its original statement of goals, especially intended for young researchers, younger applicants did very well. In this program 24.5 percent of the successful applicants were younger than 30, and 26.5 percent were from 31 to 40. And this was at a time when the portion of young researchers below 30 years in age in science in general was about 9 percent. During peer review special attention was given to proposals in which the leader of the project was younger than 35 under the principle of "all other things being equal." Therefore, the large portion of young scientists among the recipients of the ISF testifies to the high level of work of young researchers. The success of young researchers in the long-term grants program is all the more remarkable when one remembers that important factors for selection were the number and quality of previous publications and past experience, advantages particularly difficult for young scientists to possess.

Another strategy for supporting young scientists was pursued by NWO. An early evaluation of its cooperative research projects revealed that few young people participated (on average, in the period 1993–1996 a project with ten participants would contain three young scientists or graduate students). Therefore, NWO decided to overtly favor participation of young people in cooperative research projects. As a result, in 1997–2000 the share of young scientists in such projects grew to 40 percent, and in 2001–2003 it increased to almost 50 percent.

One of the authors conducted a sociological survey in October–November 2001 in twenty-one regions of Russia that yielded information about young scientists in research supported by foreign foundations.[2] The survey included 1,650 graduate students and young scientists in Russian universities, working and studying in many different specialties. In addition twenty-five focus interviews were conducted with young scientists and graduate students in Moscow.

This research showed that the participation of young people in research financed by the foundations was, as might be expected, in groups of scientists of varied ages; the young people rarely submitted proposals on their own or to programs especially directed at young people. The survey further showed that the attitude of young people to foreign support was moderately positive; their major criticism was that the amount of money available was too modest.

The survey revealed not only that young people were favorably disposed toward the grant system and foreign foundations, but also that grants had become a natural part of scientific life. As a result, the justification of such foundations and grants was not questioned. The main subject of discussion was the specific conditions under which one could receive a grant and what exactly one could do with a grant.

Thus, the material situation of young scientists improved thanks to grants. However, as the statistics on scientific personnel show, a stable inflow of young people into science has not yet occurred. Furthermore, the desire to emigrate remains high among young people. Young scientists still leave the RECs (in the BRHE program) even though their grants from CRDF are continuing. All this shows that a temporary improvement of the material situation of scientists as a result of foreign grants does not by itself lead to large systemic results. On the other hand, a survey conducted at the end of 2001 among administrators of foreign science foundations active in Russia showed that in their opinions their main contribution was support for young scientists; the administrators further believed that their foundations were helping keep young people in science. Since 53 percent of the respondents replied in this way, it appears that administrators of foreign foundations believe that their grants are having a greater impact on attracting and keeping young Russians in science than actually is the case.[3]

In other words, although the data—at least so far—do not support the view that the foreign grants are helping to keep young scientists in Russia in science, the foundations giving money do not realize this, and think they are having much greater impact on young scientists than they in fact are actually having.

Support of Women Researchers

Several foreign foundations working in Russia have set up special categories of grants for women researchers. Their programs consider the presence of women researchers among the members or supervisors of the applying group to be a positive factor for selection.

Paradoxically, an examination of the statistics on participation of Russian women researchers in grants financed by foreign foundations shows that on the average the percentage of women among the grant recipients is smaller than their percentage in the corresponding fields of science in Russia. Still fewer women are among the leaders of projects. (This low number of women leaders of foreign-funded projects may be a result of the relative absence of women at higher levels in Russian research institutions.) Data from a variety of sources indicate that women do not apply for grants in proportion to the number of women in their fields, but instead at a lower proportion. However, the share of women has grown over time both among individual grant recipients and also among project leaders. Furthermore, this increase of women participants is a general tendency for the majority of foundations, independent of whether the support of women is one of their announced priorities.[4]

The results of projects in which active participants were women can be examined in the CRDF (U.S. Civilian Research and Development Foundation) program for conducting cooperative research, the Cooperative Grants Program (CGP). In this program, aid was given to groups of scientists from the United States and the former Soviet Union in order to carry out cooperative R&D for periods up to two years. The average grant was $40,000 a year. At least 80 percent of the funds in each grant was directed to the expenses of scientists from the FSU. The first competition was announced in 1996, as a result of which 281 teams received a total of about $10 million.

The groups headed by women in the CGP projects achieved more significant successes in terms of number and quality of publications than average (both men and women). At the same time the women leaders of the projects emphasized not only the growing number and quality of publications and citation rates, but also the increasing prestige of their groups thanks to the grant they received. They also emphasized the exchange of ideas with Western partners, and the attainment of new scientific results thanks to the purchase of equipment and/or the possibility of working with modern instruments in U.S. laboratories. Other effects were mentioned less frequently (broadening of the circle of scientific interests, new topics of research, development of new contacts, including cooperation with the U.S. business sector, support of researchers and creation of conditions for the continuation of their work). These data indicate, perhaps surprisingly, that women-led groups do better, on average, than groups led by men. The question of prestige turned out to be particularly important for teams in which the leaders were women. For such groups, the foreign grants were significant in raising self-confidence, and in improving reputations among colleagues.

Thus, the data on the participation of women in research financed by foreign foundations presents surprises: on the one hand, they do not apply in numbers proportionate to their presence in their respective fields; on the other hand, when women do head such research groups the groups tend to outperform groups headed by men. This phenomenon deserves more research. It may say something important about the behavior, incentives, and personal situations of female researchers.

Support of Science in the Regions

The support of regional science (meaning, in this context, science outside Moscow and St. Petersburg) is an important priority of foreign foundations. Some foundation administrators say that this "pro–geographic distribution" position is based on the belief that giving grants throughout the territory of the country is a fulfillment of a democratic approach to the support of scientists. Others say it is not so much a question of democracy as it is of fairness; they believe that talent is more evenly distributed in Russia than past patterns of support would indicate. They further suspect that scientists in Moscow and St. Petersburg favor their own in evaluations and support, especially in money given out by central government foundations like the RFBR. The distribution principle pursued by many foreign foundations in Russia is, according to this body of thought, a way of breaking up the "big city monopoly" and allowing scientists who are just as qualified as the ones from the capital cities to obtain grants and to show their worth.

Of course it is inevitable that as more scientists from local regions receive money, those from Moscow and St. Petersburg receive less. There is, after all, only a finite amount of money available. So the distribution principle remains controversial, with some critics maintaining that science is in principle elitist, favoring the talented few, and that one should frankly recognize that the large cities have the larger part of this elite.

The promotion of a geographic distribution principle is one thing; actually bringing about wider distribution of grants is another. In all programs the Moscow region and the Petersburg region still lead prominently in the number of awarded grants: the share of these two cities amounts to 75 percent of grants awarded by the ISF, about 60–70 percent of the grants given by CRDF and INTAS, and about 60 percent of those awarded by the NWO.

In this regard the European and the U.S. programs have similar results; furthermore, the level of support of regional scientists is practically the same in grants given by both native Russian and foreign foundations. Among the grant recipients of Russian science foundations, the share of groups from scientific organizations and universities in Moscow is about 60 percent of awarded grants. Groups from St. Petersburg usually receive about 10–12 percent of grants. Of course, this does not prove that the foundations have not made a difference in distribution patterns. After all, the Russian government science foundations are also a new phenomenon, and may distribute funds more widely than the old top-down system of Soviet times.

These statistics give indirect evidence that, despite declared hopes for greater regional distribution, in the final analysis foundations recognize reality: the majority of scientists in Russia are in the large cities. Frustrating as this may be to proponents of greater distribution, it also probably indicates that the system works with considerable integrity, rating most highly the best proposals. In 2002 the share of the Moscow region combined with the Petersburg and Leningrad areas of all scientific researchers was about 58 percent.[5] And this share has not changed much: In 1997 the corresponding indicator was equal to 59 percent.

An analysis of the distribution of grants in different programs of the foundations shows that in those competitions primarily targeted at financing applied research, the share of grants going to the capital cities is even greater than in the programs on fundamental research.[6] This can probably be explained by the fact, first of all, that there are stronger ties with foreign partners in fundamental science even in the regions, and second, that the number of regions with a high level of fundamental research is greater than the number of regions with a high level of applied work.

As a whole the evidence shows that foreign foundations are not pursuing a strongly administered and regulated pro–geographic distribution policy in the grant competitions (both individual and cooperative). A comparison of the shares of projects from the regions and those of the capital cities does not bear out the existence of special support for the regions. The large percentage of successful proposals in the central region is probably because of the higher quality of proposals submitted there, although some possibility of favoritism remains.

To say that the foundations pursue a special pro-region policy seems possible only in those cases where the capital cities were excluded from certain competitions. That actually happened in a series of programs, regulated by special rules. Thus, for example, in the competition for the CRDF program Basic Research and Higher Education (BRHE, discussed in chapter 7) at first only regional universities were permitted to compete. And the most successful of the submitted initiatives were those in which regional authorities showed interest and gave support.

One particularly productive approach to supporting science in the regions is the creation of centers for shared use of equipment. Another is programs promoting access by regional scientists to central Russian and international information resources, usually through the internet. Since a large portion of the equipment base of Russian science is obsolescent, especially in the regions, such programs are effective ways of reducing regional disproportion. Such efforts were made by George Soros in the 1990s and are being continued by some other organizations today.

Support of Equipment and Facilities for Science

The effectiveness of programs for supporting the material base of science (instruments, equipment, facilities) can be evaluated on the basis of one of the programs of CRDF, the Program for Regional Experimental Science Centers (RESC). This program is oriented toward large regional S&T centers in need of modern laboratory equipment. Even in the best-equipped organizations—those in the Russian Academy of Sciences, where the material situation has always been better than average—more than 50 percent of the equipment of scientific laboratories has been in service more than ten years.[7]

Each Center received a grant of approximately $350,000 for the purchase of unique scientific equipment. Further modernization and repair of equipment was accomplished on the basis of mini-grants ($20,000 each). In Russia the program was financed cooperatively by the Ministry of Industry, Science, and Technology and local governments. In all, twenty Centers were created in this program.

In June 2005 one of the authors conducted a small survey of the Centers created within the framework of this program. It showed that one could distinguish two types of Centers: those attempting to become independent by developing alternative financial sources (e.g., industrial contacts and fees for use of equipment), and those trying to remain concentrated on their own scientific goals and existing on grants from the central and local governments. Centers in the first category dominated—they were about 62 percent of the total. In several cases the Centers that followed the second strategy—exclusively to remain science centers—did so because their equipment would have been difficult to use for commercial purposes. As a rule, organizations that had earlier been able to obtain money from the central or local governments preferred to continue this style of work rather than look for industrial support. They were constantly in search of government money. The others were equally busy in attempting to broaden commercial activities.

Even among the Centers that were overtly commercially oriented there were differences. In this group one can see two subgroups: centers trying to diversify the sources of financing and also their activities (research, education, innovation, commerce); and those trying to convert themselves entirely into commercial structures rendering services to various types of users.

The Centers were asked what factors they consider necessary for their sustainable functioning in the future. They answered that they were trying to become sustainable in several ways: developing more commercial ties (15 Centers, or 71.4 percent), diversifying their activities (6 Centers, or 28.6 percent), seeking foundation financing (4 Centers, or 19 percent), and seeking government financing (3 Centers, or 14.3 percent).

The survey shows the importance of capable leaders who seek the right mixture of sources for sustaining their Centers. Choosing good managers is probably more difficult than choosing good scientists, since formal criteria for selection of managers are much less clear than for scientists.

The program amply demonstrates that giving money is insufficient to obtain the effective use of the equipment that is purchased with it. Material support needs to be accompanied with the launching of research, education, innovation and other kinds of projects and initiatives that create the demands for the equipment. In this case these demands were stimulated by another initiative of CRDF, the BRHE program.

Results of the Partnership Function of Foundations

Effective partnerships mean obtaining mutually beneficial results by researchers of different countries on the basis of cooperative research projects.

An evaluation of the results of programs of cooperative research carried out by NWO, IREX, CRDF, and INTAS shows that partnership in science is truly profitable, both to Russian and to foreign researchers. In the CRDF cooperative research program, Russian grant recipients on completion of projects noted that they consider the chief advantage of work on cooperative projects with the United States to be conducting research on the equipment of their U.S. partners; in this way they mastered new methods of research, especially in experimental areas, and also obtained new information. A third of all participants in the projects emphasized these benefits. Second in significance was the exchange of ideas and mutually supplementing knowledge during the U.S.-Russian cooperation. In 1998 this was cited by almost a fifth of all grant recipients. It should also be noted that in long-term projects the significance of exchanges of ideas and expertise grows. According to evaluations of projects completed in 2002, almost 90 percent of the grant recipients cited such growth.[8]

Evaluations of completed projects carried out by NWO gave similar results: For the Russian side the access to the scientific equipment of the foreign partners was important, but so was the appearance of new scientific ideas and the mutually supplementing expertise of the participants.[9] On the IREX program in the humanities and social sciences the chief advantages named were cooperative work, professional development, and access to scholarly infrastructure.[10]

Three years after completion of projects, former grant recipients of CRDF were given the same question about the main advantages of carrying out cooperative projects. It turned out that with time the views of the participants in cooperative programs changed somewhat. As time passed, a growing—even predominant—number of respondents named as most important the development of scientific careers, international recognition, growing reputation in the scientific community, the development of new ties not only with U.S. but also with European colleagues, the growth in the number of publications, and also broadening of knowledge and the posing of new scientific tasks. Of course these achievements can not be attributed only to receiving CRDF grants, but the fact that the participants gave such a high evaluation to long-term consequences of grant support is, in itself, remarkable. Such results show that as time goes on, Russian participants in cooperative projects take a less instrumental view (access to foreign equipment) and a broader, more comprehensive one (improvement of scientific careers and better research).

Immediately on completion of projects about one-fourth of the members of the Russian team expressed the intention to work on further commercialization of the results achieved. In three years 23.7 percent of the respondents noted some progress in commercializing the results (such as submitting patent applications, developing prototypes, conducting marketing, launching pilot plant production). The largest number of commercially significant results was achieved in the biological sciences, bioengineering, and chemistry. With each new competition of CGP (the Cooperative Grants Program), Russian attention to commercialization grew. Thus, in the program completed in 2002, 31 percent of the Russian project leaders noted that they were preparing for commercialization of the project results. In this way, the partnership function of the foundations gradually changed into a commercial function. An analogous tendency could be seen in the programs of cooperative projects of INTAS: there, as a result of cooperative research, 550 patents were received. Such progress stimulated a new program of innovation grants.

The Russian members of the teams were not the only ones to benefit. Western scientists obtained interesting results and unpublished information through cooperation with Russia, and gained access to unique regions (especially important in areas like zoology, botany, and the earth sciences).

An evaluation of INTAS projects showed that in cooperative projects the financing by the foreign partner is often small (25 percent); for many European scientists such cooperation is a "test run" to see if it is possible to work with Russian partners on larger projects.[11] The Russians made a definite intellectual contribution, since the ideas for the projects came most often from the Russian side. Over time the number of member countries of INTAS constantly increased, demonstrating that for the countries of the European Union this initiative was interesting. The importance of cooperative projects was emphasized by the Alexander von Humboldt Foundation.[12]

The chief result of partnership projects is scientific publications; according to this indicator the programs of cooperative projects have been highly evaluated by such organizations and foundations as INTAS, NWO, CRDF, and the Fulbright Program. The evaluations of projects of NWO show that the number of publications of Dutch scientists in refereed journals grew thanks to cooperation with Russian colleagues, just as the publication activity of project participants grew as a whole. In 1997 there was an average of four publications per project, and in 2002 that indicator grew to eight publications per project.[13] The level of publication activity on INTAS projects turned out to be even higher: on average there were about ten publications per project.[14] The Fulbright program showed that as a result of cooperation of Russian and U.S. scientists the number of cooperative projects funded by different U.S. foundations grew.

An indicator of the success and importance of cooperation for both sides is the share of scientific groups continuing joint research after the completion of support from the foundations. In the programs of NWO and CRDF this indicator is on average 95–97 percent. An analysis of the results of IREX programs gives a similar result: 94 percent. Furthermore, 85 percent continue cooperation at the institute level, not just individual scientists.[15] An analysis of the INTAS projects also confirmed that a majority of partners continue cooperation.[16]

Results of Support of Commercialization of R&D

Foundation interest in commercially significant projects did not arise immediately. It came only after potential commercial results were obtained in programs that did not have practical goals. These results showed that the work of Russian scientists can be used profitably by all participants, including the side giving the support.

The commercially oriented programs with the longest histories are those of the ISTC and CRDF. Beginning in 1997 the ISTC supported partnership projects directed toward the commercialization of R&D. Partners financing the R&D of Russian researchers can be foreign government agencies as well as private organizations and firms. The idea was to develop personal ties between Russian scientists and foreign organizations interested in their research and also to foster commercialization habits among Russian researchers.

To help Western organizations and firms in their search for scientific groups in the countries of the CIS, the ISTC created a special data bank containing résumés of promising Russian research. By the beginning of 2003 the database contained 1,800 such résumés. The ISTC was thus beginning to fulfill not only consultative functions but also middleman functions, since at the request of Western organizations it identified Russian groups for the fulfillment of concrete tasks.

Among the Western partners were such companies as 3M, Lockheed Martin, Bayer AG, Shell, Hitachi, Mitsubishi, and Samsung Electronics. Government agency partners included the U.S. departments of Defense, Energy, and Agriculture, and the Environmental Protection Agency (EPA). Russian scientists also actively worked with foreign national and international laboratories such as Los Alamos, Lawrence-Livermore, CERN, and DESY (Deutsches Elektronen-Synchrotron). Before beginning such cooperation, the partners were required to give the ISTC assurance that the character of the requested work served the goals of conversion.

The program showed that the research of Russian scientists is valued in the West. Evidence for this is the constant growth of financing by the Western partners, from $1 million U.S. in 1997 to $207 million in 2005.[17] By 2003 the financing from partners equaled the financing given out by the ISTC itself, and it continued to grow in 2004. The growth in financing of projects by the private sector indirectly speaks of the success of this initiative. According to statistics for 2004, on the basis of partnership projects five commercialization agreements were made, which led to the creation of three hundred new employee positions for former ISTC grant recipients. More complete quantitative statistics showing the results of the partnership programs do not exist, since the ISTC plays only a helping role in choosing projects and drawing up contracts. The basic conditions and results of the work are determined by the partner organizations.

An analysis of the types of partnership projects (see table 8.2) shows that foreign companies and organizations are particularly interested in biotechnology, earth sciences, physics, and new materials. These same areas are, of course, actively developing throughout the world.

Looking back over the evolution of the ISTC we can see that there has been an important change in the basic motivations behind it. The ISTC program was established primarily for the diversion of Russian military researchers to civilian purposes. As time went on, and as it became more and more clear that such a goal cannot be accomplished at a time when both the Russian and U.S. governments are supporting much military research, the reason behind the ISTC shifted to commercially profitable joint research by Russian and foreign scientists. Many Western companies learned that it was profitable for them to cooperate with talented (and relatively inexpensive) Russian scientists. This new interest in mutually beneficial cooperation in science and technology explains the fact that the budget of the ISTC continues to expand even though the original goals of nonproliferation and diversion into civilian research are no longer nearly as important as they originally were.

According to a 2001 survey, 47 percent of the foundations actively working in Russia believe that their chief contribution is in aid to Russian scientists in commercialization of their products, an increase from the past. Currently a

Table 8.2. Basic topics financed by partnership projects of the ISTC
(statistics for January 1, 2003, for countries of the CIS)

Areas of research	Share of financing, %
Biotechnology and earth sciences	41.9
New materials	11.3
Physics	5.6
Nuclear reactors	4.7
Ecology	4.6
Non-nuclear energy	4.4
Production technologies	3.6
Chemistry	2.5
Instrument construction	2.4
Information & communication technologies	2.1
Space, aviation, land transport	1.0
Other	15.9
Total	100

Source: ISTC. Partner Brochure, 2003, p. 11.

basic problem is that the foundations give out rather modest sums for commercialization.

Problems of Intellectual Property and Foreign Foundations

Despite the fact that foreign organizations and foundations finance R&D in the pre-competitive stage, the result can be an invention that is patentable and therefore deserving of protection. The Russian market is still weakly interested in technologies and products of high scientific content. Therefore rules regulating the rights to intellectual property (IP) are one of the most important mechanisms determining the possibilities for profits from commercialization.

Most foreign foundations, especially those that support scientific exchanges and give grants primarily in non–applied research areas, do not give significant attention to problems of IP. As a rule such organizations and foundations leave the rights to IP to the performers of the projects (the grant recipients) and do not themselves ask for any property rights. On the other hand, organizations and foundations oriented toward carrying out partnership and innovation projects (first of all, the ISTC, INTAS, CRDF) have special policies on IP. The most liberal toward the Russian side is the CRDF.

The CRDF governs its activities in accordance with the Russian-American Agreement of 1993, and also relies on the experience of two other U.S. foundations—the U.S. National Science Foundation and the ISF. The CRDF does not make any claims to the intellectual results obtained as a result of its programs.

The CRDF worked out recommendations for the sharing of rights to IP among performers of projects, according to which all IP in cooperative research belongs to the group that created it. The same is valid regardless of whether the IP came from the Russian or the U.S. side. If IP was jointly created as a result of cooperative work, then its regulation is a subject of agreement in each concrete project, but both sides must be co-owners of such IP. In any case the distribution of rights to IP created in Russia is in accordance with current Russian legislation. As a rule a project is not approved until a mutually acceptable agreement about IP has been reached. The clarity of the procedures and the transparency of relations on IP are attractive factors for potential investors, i.e., industries and small businesses.

The ISTC developed regulations of IP rights at the very beginning of its activity (the middle of the 1990s), when the Center promoted only conversion projects. At that time the norms adopted were entirely adequate. Later, however, the situation changed: partnership projects appeared, and the Russian side began to take up questions of IP more fully. Therefore, the mechanism of regulating IP rights in the ISTC needs modernizing.

For the ISTC, rights to IP are regulated by article 13 of its charter. According to this article each side (the financing side and the performing side) and the ISTC have rights to non-exclusive, irrevocable, free license with the right of sublicensing in all countries for the transfer, reproduction, and broad distribution of S&T journal articles, reports, and books directly connected with the project. If IP is created, then the performer of the work (the Russian side) gives the financing side exclusive, irrevocable, free license (with rights to sublicensing) for commercial goals on the territory of the country financing the project. The sides also agree about compensation for authors and inventors. If a project is financed by several countries, there is agreement in each case in accordance with the laws and legal norms of the participant countries. The Russian performers can commercialize the work in Russia, but not in countries that financed the work. However, since the foreign organization financing the work is usually a government agency, the disposition of IP rights still sometimes remains obscure. According to 2004 data, as a result of ISTC projects 240 inventions were created, 162 patent applications were submitted (of which 81 were in Russia), and 31 patents were received (all in Russia).

In partnership projects the situation is even more complex. Special agreements are usually made in each particular instance. However, such an agreement must be signed by three sides—the partner, the Russian performer, and the ISTC. Partners sometimes demand equal rights to IP if they are financing the project. At the same time there are more and more agreements in which the Russian side may receive rights to IP in areas outside the interests of the partners.[18]

Before a project begins, the Russian side may have clearly delineated its interests in already existing IP, but this is not always the case, and problems can result from this oversight.[19] A number of organizations (ISTC, CRDF, NWO) have initiated seminars and training sessions to try to anticipate such issues. The ISTC has a legal department that helps Russian participants on issues of IP, and also prepares them for negotiations with foreign partners.

In the early nineties Russians often made mistakes based on inexperience when they signed contracts with foreign partners.[20] They often neglected problems arising out of commercialization of created IP, patent applications, confidentiality, and mechanisms for resolving disputes. Today such mistakes are much less common, in part because of assistance and consultation on the part of foundation staffs.

Results of the Security Function

Stimulating Conversion in Defense Science

The investments of Western organizations in conversion have been very significant. The largest international organization supporting conversion of former defense scientists is the ISTC. This center established the goal of reorienting scientists from the former defense sector to civilian needs by means of S&T projects.

The ISTC invited proposals for S&T projects from Russian institutes and financed and supported groups of scientists. A requirement was that in such work no less than 50 percent of the participants must be former defense scientists. Groups worked in collaboration with Western organizations from the country or countries financing the projects. The collaborating partners did not participate in the work but instead served as consultants facilitating international ties, something important for scientists who earlier worked in closed organizations and cities (secret defense sites). The collaborators guaranteed that the goals of the project aided conversion and that the results found application in the solution of current problems of fundamental and applied science.

The chief financing sides are the United States, the countries of the European Union, Japan and the Republic of Korea, and since 2004 also Canada, which became a member of the ISTC. In 2004 more than 27,000 scientists from the CIS (including more than 21,500 from Russia) received support for ISTC projects.[21] The total amount of financing was $47.3 million U.S., and the average project cost was $350,000.

In Russia the chief recipients of funds are scientific groups from organizations located in Moscow and the Moscow region, Nizhnyi Novgorod, St. Petersburg and its surrounding area, Cheliabinsk, and Novosibirsk. The distribution of financing by topic shows that priorities at the present are biotechnology (38.5 percent of funds), ecology (12.1 percent), physics (10.7 percent), and chemistry (9.9 percent). The priorities have changed over time; in 2002 chemistry was not among the priority groups and second after biotechnology was production technology (12.3 percent). A general picture of the distribution of financing is shown in table 8.3.

Table 8.3. Basic topics financed in ISTC projects (for countries of the CIS in 2004)

Area of research	Funding, in U.S. dollars	Number of projects
Biotechnology and the life sciences	21,545,263	61
Ecology	6,776,950	23
Physics	5,982,979	23
Chemistry	5,474,542	17
Nuclear reactors	4,694,106	19
New materials	3,107,877	13
Information and communication technologies	1,900,293	7
Instrument construction	1,837,598	7
Non-nuclear energy	1,722,214	7
Space, aviation, land transport	1,531,012	6
Production technologies	859,867	4
Fuel elements	156,231	2
Other	373,500	1
Total	55,962,437	193

Source: ISTC Annual Report—2004.

The projects demonstrated that on some commercially significant projects the results were so promising that successful small innovation enterprises were created on their basis. Unquestionably, this was a contribution to conversion.

From the point of view of preventing brain drain, the results of ISTC work illustrate that the level of emigration on ISTC projects is not high. On the other hand, a survey of 602 former ISTC grant recipients conducted in 2002 by U.S. sociologists shows that about 20 percent of these Russian scientists do not exclude the possibility of work in such countries as North Korea, Iran, Iraq, and Lebanon.[22] However, the programs of the ISTC lower the probability of the outflow of scientists to such countries. This is confirmed by comparing the intentions of scientists participating and not participating in ISTC projects. Among those who were not participants 28.1 percent expressed the desire to go abroad, but among those on ISTC projects only 14.9 percent.

However, one should notice that the emigration of scientists, even when it happens, is usually not to third world countries; instead, scientists as a rule emigrate to the countries where they have ISTC partners—to the United States, the European Union, and Japan. The leadership of the ISTC does not consider such emigration to be dangerous.

The U.S. CRDF has also made an attempt to evaluate its conversion programs. In 2004 an external group of consultants made an evaluation in which one of the questions asked of former grant recipients was: "To what degree does CRDF help former defense scientists reorient to civilian projects?" The results show that about two-thirds of the respondents think that the contribution of

CRDF to conversion was quite perceptible. The remaining third thought that CRDF was not in a position to influence the process of conversion significantly if this goes against the policy of the Russian government.[23]

Thus the success of conversion depends not so much on the activity of foreign foundations as on the priorities of the Russian government. Receiving grants for carrying out civilian projects, scientists who had earlier worked in defense reorient themselves, but this reorientation is temporary. When government defense orders reappear, the researchers participate in them. Full conversion occurs only in cases where the organization does not receive defense contracts. Since in the last two or three years the volume of defense contracts in Russian science has approximately doubled, it is not likely that in the near term the full reorientation of researchers to civilian purposes wished for by Western sponsors will occur.

In conditions of the growth of government financing of defense R&D, the earmarking of funds of foreign foundations for conversion becomes pointless. Therefore in part the ISTC has refocused on carrying out partnership programs, that is, on the fulfillment of the commercial function, which, on the one hand, is needed in Russia at the present time, and, on the other, if successful is simultaneously a contribution to conversion.

The Prevention of Brain Drain

Brain drain is related to conversion, but is obviously not the same. Indeed, as already noted, almost all brain drain is to Western countries, not to "rogue nations." An examination of the influence of foundations on brain drain reveals both what measures foundations have taken in an attempt to stem brain drain and also what actual influences on brain drain they have had.

An evaluation of brain drain shows that there exists a rather strong relationship between the intensity of emigration of scientists and grant financing. Much depends on the type of grant offered (for work abroad, or for conducting research in Russia); the type of research being supported abroad (fundamental or applied); and the size and duration of the grant.

Research abroad is often the first step to actual emigration. On the other hand, financing research in Russia helps scientists to remain in their institutes and continue their scientific work. Other factors that can work against outflow are long-term grants in Russia (no less than 2–3 years) and relatively large grants. Such factors help scientists devote themselves totally to their research and not be distracted by the search for additional sources of funding, including those abroad.

During research abroad, much depends on the character of the work being conducted and the length of time spent abroad. The longer that period, the greater the probability that the scientist will not return to Russia. This is illustrated, at least partially, by the statistics on grant recipients from the NSF. A long stay by Russian scientists in the United States increases the likelihood of emigration. After emigration a large number submit applications to NSF for cooperation with colleagues remaining in Russia.[24]

Conducting applied research also leads to growing emigration—scientists remain abroad in order to continue their projects. Furthermore, finding work in industrial laboratories and foreign companies is, as a rule, easier than in the academic sector.

The regulations enforced by foundations also have a great influence on the intensity of brain drain. Foundations such as INTAS, CRDF, and IREX have established limits to the amount of time Russian scientists can stay abroad on their grants. According to the rules of CRDF, Russian scientists can be located in the United States in total no more than three months each year. The Wellcome Trust limits the time of residence of Russian scientists in the UK in the same way. Such rules diminish brain drain.

Nonetheless, an evaluation of CRDF programs shows that success in preventing brain drain is elusive. On completion of work in the cooperative research program, 18 percent of the project leaders noted that someone on their teams went to work abroad for a period of one-half year or longer. A fifth of those leaving were undergraduate and graduate students. In three years, according to a repeat survey, it was learned that in 44 percent of research teams somebody left to work abroad for a year or longer, and 60 percent of those leaving were young scientists. However, one should note that it remains unknown what share of those who left returned later. Evidence from other research indicates that, as rule, the portion who return is not more than 10–20 percent of those who left.

Such a conclusion is supported by the results of still another initiative, a foundation established by French mathematicians called Pro Mathematica. One of the key goals of Pro Mathematica was prevention of brain drain among mathematicians in Russia. Pro Mathematica is a non-governmental organization created in 1992 in reaction to the needy position in which Russian mathematicians found themselves after the collapse of the USSR. Funds for the support of Russian mathematicians were collected from the following sources: 85 percent from government agencies (French ministries of Education, of Science, of Defense), 14 percent from various scientific societies, and 1 percent from other sources. The named ministries gave about 100,000 francs a year over the course of an almost seven-year period. In 1992–1993 financing by these grants equaled $40 a month, and grants were awarded for a half-year. About one hundred scientists a year received such grants, primarily young people (up to thirty-five). Then the grants were increased to $50 a month and were awarded annually.

What kind of results were achieved? The Russian mathematical school was preserved, but a significant part of it was transferred abroad. The program supported about fifty graduate students, but many of them left for France or the United States. An analysis of the current location of former grant recipients of Pro Mathematica shows that 37 percent have remained in Russia and are working there, 1.2 percent have remained in Russia but are not working in their specialties, and 18.5 percent have gone abroad and are continuing their careers in mathematics. The fates of the remaining 43.2 percent are unknown.[25] It cannot be excluded that some of these people also are working abroad but not in mathematics.

Thus we see that the programs of foreign foundations are not able to stop brain drain, but they can hold it back during the period of the grants. A survey of 106 former grant recipients of INTAS demonstrated the same effect; the program retarded brain drain during the period of the grants but could not halt it.[26] It should also be noted that surveys of former grant recipients show that a small group of respondents think that the programs of foreign foundations actually stimulate brain drain, whatever their intentions. A recent survey of former grant recipients of the Alexander von Humboldt Foundation shows that 13 percent of the respondents held the opinion that foreign foundations stimulate emigration of scientists from Russia, and 55 percent believe that the influence of the foundations is of a dual nature: they both stimulate the outflow and at the same time strengthen the motivations of some scientists to stay in Russia.[27] It is thought that the stimulation of outflow happens particularly among young people who go abroad for study or participation in conferences; there they begin to understand better their ability to compete and then sometimes make a choice in favor of comfortable foreign laboratories. For mature scientists emigration abroad is less attractive.

On the other hand, an evaluation of the CRDF programs indicated that there are no connections between the intensity of emigration of scientists abroad and international travel. In a number of cases the scientists who emigrated worked on projects for which there were no foreign trips supported by the Foundation; furthermore, in projects in which much travel was involved the brain drain was the least.[28] Russian scientists who feel free to travel back and forth often see no reason to emigrate if they have good jobs at home.

The most successful example of the prevention of brain drain was a local project carried out in 1995–2000 in Nizhnyi Novgorod called the International Foundation-Center for Promising Research. The center had a small budget (during the period of its work it was given less than $500,000), but even this modest financing led to clearly positive consequences as a result of an original concept incorporated into its work. Understanding that grants for graduate work and conducting research abroad stimulate the outflow of personnel from Russia, it was decided to create the opposite—that is, to invite scientists from abroad for cooperative work in Russia and to pay for these visits.

The center had two basic directions of research: the competitive support of cooperative research projects (up to a half-year in length) and also visits of foreign and Russian specialists to give lectures to undergraduates, graduate students, and researchers in Nizhnyi Novgorod. During peer review of the projects, in addition to standard criteria the prospects for cooperation were taken into account. In the opinion of the foreign participants the project was quite successful. Here was an unusual situation in which the Russian side was not the seeker but the financer of cooperation, although also with the participation of foreign philanthropic funds.

The results of these projects were unprecedented: none of the young researchers—and they were 63 percent of all participants—emigrated abroad. More than half of the researchers continued contacts and cooperative work with their Western colleagues, sometimes on the basis of foreign invitations and grants. Several institutional partnerships with foreign universities were also created.[29]

Thus, the problem of brain drain is complex. Grants to Russian scientists to travel and work abroad do not necessarily increase brain drain. Much depends on the details: How long will the stay abroad be? What is the nature of the work? Is there free flow back and forth? Do scientists travel in both directions?

Results of the Reform Function

There exist a series of programs of foreign organizations and foundations supporting organizational and economic means to spur innovation, what is often called innovation infrastructure. Among the most active organizations and foundations supporting innovation infrastructure are the Eurasia Foundation, the British Council, CRDF, and TACIS.

Initiatives of the Eurasia Foundation and the British Council for Creating Innovation Infrastructure

In the period 1995–2004 the Eurasia Foundation gave out fifty-seven grants of $35,000 U.S. each to stimulate innovation. The total amount of financing was $1.5 million. Projects were launched in sixteen regions of Russia to develop consulting in innovation management, to train personnel for small innovative enterprises, to develop financial mechanisms for innovation, and to form innovation infrastructures.

The majority of projects (60 percent) were concentrated in training and consulting. In total forty-four organizations were supported, including Technology Parks, business incubators, innovation and educational centers and institutions, commercial and consulting organizations, and agencies for support of business undertakings.

An evaluation of the long-term results of the implementation of many of these projects was conducted in the spring of 2005. It determined that participation in the program raised the reputations of the grant recipients, helped employees realize their professional aspirations, and broadened the client base and the spectrum of services offered. In educational efforts specializing in the preparation of managers for innovation business, the number of attendees in courses grew by several times, and new specializations appeared.

A little more than half (52 percent) of the polled Eurasia Foundation grant recipients said that their efforts to stimulate innovation had been copied by others, particularly by creating Technology Parks and incubators. In particular, the model of a student incubator that the Foundation supported was considered a promising initiative from the point of view of developing and spreading an innovation culture.

At the same time, in a number of projects the planned results were not attained. Scientific research institutes inherited from Soviet times often were unreceptive to efforts to add new innovation structures. Single efforts to do so were often not adopted or adapted by others.

One more reason for the inadequate effectiveness of infrastructure projects was that they required long-term financial support. On the basis of a grant from the Eurasia Foundation, which on average did not exceed $35,000, it was possible to conduct effective training and to give consulting services, but not to create new organizational structures. In the opinion of former grant recipients and experts, in order for an infrastructure project to achieve a result, the financing must be increased to at least $100,000 or $200,000, and the length of the project must be no less than two or three years. One can not expect returns from innovation projects lasting only a short time.

Infrastructure projects were the most effective when local government administrators stepped forward to help. Evaluation also showed that the productivity of projects was higher in those organizations that had a clear specialization. The support of organizations having significant income from commercial activity turned out not to be so effective; the project receiving special support tended to "sink" among commercial projects and have only a short-term effect. The leaders of commercial organizations rarely attempted to use the gains made in the grant project in their other initiatives.

The results of the support of innovation projects by the Eurasian Foundation allows one to draw the conclusion that the most effective ones are those combining training and later consulting by small innovation firms. Support must be continuous and resourceful in order to achieve success. The copying of successful models makes sense after positive examples have been identified on the regional level; it then becomes possible to form "lobby groups" for the representatives of innovation business.

The British Council in 1996–2003 gave to the support of science and technology in Russia about £2.5 million. The average annual expenditures consisted of £200,000 directed at implementing projects and £230,000 at information resources. After the middle of 2004 the activity of the British Council in the support of the S&T sphere in Russia was transferred to the Ministry of Trade and Industry of the UK.

A special characteristic of the activity of the British Council was that it did not have a program supporting research in the traditional sense, and also did not finance organizations promoting the transfer of technology; these latter functions were left to British companies. In this way the British Council attempted to avoid being a substitute for private capital. It set its basic goal as the implementation of pilot projects containing new mechanisms for the organization of S&T activity, and, correspondingly, helping reform science in Russia. The British Council implemented its initiatives in close cooperation with Russian government agencies. The time length of projects was on average 1–2 years, and did not exceed three years.

The support of infrastructures of innovation activity by the British Council resulted in the creation of a center for the commercialization of S&T developments of the Institute of Chemical Physics of the Russian Academy of Sciences in Chernogolovka, and implementation of a pilot program for Russian-British industrial cooperation in the Volga region.

The Center for Commercialization came about as a result of a decision taken at a joint meeting in 2001 of the British Council and the Academy of Sciences. The Center, a pilot project in the Institute of Chemical Physics of RAN, was also partially supported by a grant from the Eurasia Foundation. The basic goal of the Center was organizing the commercialization of the results of R&D and implementing an innovation policy in the Institute. In the framework of this project the British Council financed the training of personnel, the preparation of methodological materials, and the conducting of seminars. This Center, which was essentially a technology transfer office, conducted technology audits directed at choosing technologies with commercial potential. It also developed norms for preserving commercial secrets and for distributing license payments.

Technology audits demonstrated that the Institute of Chemical Physics was producing interesting results, but was not adequately marketing potential technologies. The British Council gave help by searching for a middleman firm, settling on British Gryphon Emerging Markets Ltd. On the basis of a contract this firm was presented with eighteen short descriptions of developments for which over six months partners in commercialization were sought.[30] Interest was shown in eleven of them.

In the course of this project intellectual property rights were strengthened, and a method was worked out to distribute licensing payments. Payments coming to the Institute were distributed in the following way: up to 55 percent was paid to the authors as rewards, 20 percent to the department in which the licensed object was developed, and 25 percent to the Institute. Despite this progress, questions of regulating rights to intellectual property created in the Institute were still not entirely solved in view of the inadequacies of the existing legislation. Because of contradictions in the legal basis and in the lack of preparation of institutes for commercialization, the model was not duplicated in other scientific organizations of the Russian Academy of Sciences. It had been hoped that similar centers would be created in forty other institutes of the Academy, but this has not happened.

The British Council also supported a program on industrial cooperation, which introduced a scheme, well known in the UK, of training youth for work in small innovation businesses (Teaching Company Scheme—TCS). The idea of this program was to connect small businesses and science education institutions. Undergraduate and graduate science students were seen as the connecting element; they would fulfill short-term R&D projects in a small firm, and their work would be supervised both by the university or institute and by the firm. In that way an educational result also would be achieved, since personnel were prepared for innovation activity. At present a similar initiative is being continued by the Russian Foundation for Assisting the Development of Small Enterprises in S&T.[31]

The TACIS Program for the Development
of Regional Information Infrastructure

In 1999 TACIS, together with the Ministry of Industry, Science, and Technology, implemented a program called Innovation Centers and Science Cities in the "pilot regions" of Obninsk, Reutov, Troitsk and Kol'tsovo. Later Kol'tsovo also received the official status of Science City.

The implementation of the project followed three basic goals: developing a model Science City, creating innovation infrastructures in individual institutes to promote commercialization, and establishing ties both between Russian scientific organizations and industry and also between Russian and European innovation centers.

The project had a predominantly educational character. It included training for two groups of specialists from Russia. In the first were leaders and officials of federal agencies whose work was directly connected with developing areas of high concentration of S&T potential. The second group were mayors, vice-mayors, and leading urban authorities. The first part of the training took place in Russia, and the second in countries of the European Union during two-week study trips.[32] Seventy-seven Russians and twenty European experts took part in the project; as a result seven books were published, including a glossary of terms in the areas of science, technology, innovation, and business. During the project SWOT analysis (strengths, weaknesses, opportunities, threats) was conducted for each of the four cities. Seminars were conducted on business planning, strategic marketing, negotiations, and the evaluation of production costs.

It is difficult to evaluate adequately the long-term results of the project, since there is no information on how the training and innovation centers influenced the development of Science Cities. The available statistics about Science Cities indicate that the transition to innovative development has still not occurred. On average, in all Science Cities at present the share of innovation production is about 20 percent, and the number of large enterprises pursuing innovation activity and manufacturing new products does not seem to be growing.

The experience of the development of Science Cities indicates that the attitudes of regional authorities and the level of local management have critical significance. For researchers from institutes and scientific organizations to be able to work with business representatives a common language must be developed, something that occurs only with difficulty.

Summarizing the results so far of support for innovation infrastructures, one can conclude that the most successful initiatives are those that attract undergraduate and graduate students to consultation services involving both small enterprises and scientific research organizations. Young people are more flexible than their seniors and have advantages in finding the necessary common approach.

Forming new objects of innovation infrastructure is very complicated, and so far successful results are not conspicuous. The innovation organizations that are sustainable after the cessation of foreign financing seem to be those oriented toward specific buyers, including regional authorities. Support from regional authorities in the form of direct financing, real estate, and administrative assistance can at least partially replace the earlier foreign aid.

Bringing Research and Teaching Closer Together

As discussed in chapter 7 (Strengthening Research in Russian Universities), three U.S. foundations (The John D. and Catherine T. MacArthur Foundation, the Carnegie Corporation of New York, and the Civilian Research and Development Foundation) have made an effort to bring research and teaching closer together in Russia, concentrating on the universities as the places to achieve this linkage. These efforts have been made both in the natural sciences (the BRHE program) and the social sciences (the CASE program). The support of the Russian Ministry of Education and Science for these programs, as well as that of local governments and institutions (in total, the Russians more than matched the money devoted to these efforts by the U.S. foundations), shows that these programs have had unusual success in attracting support within Russia. This reform, therefore, deserves careful attention in the future. The traditional separation of teaching and research at the highest levels in Russia is so strong, however, that it would be very premature to pronounce this reform an unqualified success. Nonetheless, the impact of these programs has been great, and even if the traditional Russian organizational pattern remains strong, as is likely, alternatives have clearly been created. The term "research university" now has a currency and even a popularity in Russia that was totally absent as recently as ten years ago. Current reforms in the Russian Academy of Sciences point toward a growing recognition of the importance of combining teaching and research. Therefore, even if the ways in which Russia organizes its research and teaching establishment in the future differ considerably from the U.S. model of the research university (again, as seems likely), the value of the BRHE and CASE programs already seems evident.

Unintended Consequences of the Work of Foreign Foundations in Russian Science

Among the unplanned effects of foreign organizations and foundations in Russia are changes in themes of scientific research, and also changes in the methods of work of the foundations themselves under the influence of Russian conditions. Another unplanned result of the work of foreign organizations and foundations is the appearance of private Russian philanthropy.

Stratification of the Scientific Community
and Changes in Themes of Work

A growing stratification within the Russian scientific community, with the appearance of a small group of successful and prosperous scientists, is one of the unintended consequences of the work of the foundations. Just as the introduction of a market economy in Russia led to greater disparities of wealth among ordinary citizens, so also the introduction of competitive systems of grants, fostered by foreign foundations, has led to differences between "winners and losers" in the scientific world. And just as New Russians appeared in the economy as a whole, so also New Russian Scientists have emerged, those who are using the new systems to their advantage. Several different types of New Russian Scientists now exist.[33] Among the more common are those who have been recipients of grants from foreign foundations (which, as already noted, usually receive higher salaries than before) and also the Russian employees of foreign foundations, firms, and universities, including, especially, those with offices in Russia (often in Moscow). This group is currently growing in part as a result of outsourcing, especially in areas connected with computer science and the applied sciences.

Another type of privileged Russian scientist is the group who participates in "pendulum migration" and who has made this style of work permanent. The numbers involved here are not large, but especially in Moscow and St. Petersburg there are now scientists who spend on average a half-year abroad. Some have teaching or research positions in European or U.S. universities, where they annually spend one or two semesters, then returning to Russia for the remainder of the year. Other Russian scholars, especially in the humanities and social sciences, are employed as experts by various foreign government and political and social organizations. Examples here would be Russians working for foreign environmental organizations, human rights groups, and think tanks. Finally, a number of Russian scientists and engineers have become high-level managers and scientist-entrepreneurs. They enjoy living standards far above those of the average research scientist.

Not all of these privileged scientists and scholars in Russia achieved their high incomes on the basis of support from foreign programs and foundations, but many did, and this group is now rather stable. Moreover, one can see a tendency toward a growing concentration of grants in a few groups and institutes. Thus, statistics collected in 1997 and 2002 on cooperative research sponsored by NWO indicated that in 1997, 63 percent of Russian participants simultaneously had grants from other foreign organizations, while in 2002 the same statistic was 66.3 percent.[34] (One suspects that these statistics may actually be low, since not every scientist with several simultaneous grants from different organizations can be expected to report that fact.)

At the same time the incomes of scientists in Russia are relatively low, and about 70 percent of scientists consider themselves to be in the least financially secure stratum of the population, and only 30 percent in the middle stratum. The economic situation of scientists during the last five years has improved for about 30 percent, but has remained practically unchanged for a larger portion, about 40 percent. Moreover, almost 60 percent of scientists think that in the near future the level of their incomes will not improve. This picture has not substantially changed in recent years.[35]

Dependence on foreign financing may have the most negative consequences if it suddenly ends: for example, in the crisis period of 1998, payments on grants from the ISTC were stopped for two months, and this nearly led to a freezing of the work of several institutes.

Changes in Themes of Work

Still another consequence of active foreign participation in Russian science is the influence on themes of work. This influence may be direct, through the definition of topics of competition, or indirect, when one of the requirements is the presence of ties with foreign colleagues. In the latter case the supported directions correlate with themes being worked on in the countries with which cooperation is developing.

Without question, the first type of influence—the definition of topics of competition—is the most significant. This was especially apparent in the middle of the 1990s when foreign foundations actively supported several areas of social science—sociology, political science, economics, gender research, etc. The number of political scientists and specialists in gender research in Russia increased by several times. (Of course, in some instances this was more a case of "relabeling" than professional retraining; for example, many philosophers of Marxism suddenly became "political scientists.") The themes for which the foundations declared their support were, in their opinion, of high priority and important, even though they may not have always been the most timely or culturally acceptable in Russia. In this instance there was adaptation to the priorities of this or that foundation, but not always a genuine development of the research direction.[36]

Another direct effect of foundations' activity flows from their emphasis on conversion of military research to civilian purposes. This emphasis has been so strong in the work of a few foundations—especially the ISTC—that it has caused both complaint and some dishonesty. By offering grants only to, or primarily to, scientists who were earlier engaged in Soviet military research, these foundations have increased the temptation for scientists to exaggerate the degree to which they were involved in such defense work in order to improve their eligibility for grants. Other more honest scientists have noticed this and resented the fact that because of their integrity they have been excluded from financial support. We have heard complaints from scientists in Russia of the following type: "In the days of the old Soviet Union I always refused to get involved with military research because of my disagreements with the Soviet regime, even though I would have been more highly rewarded if I had gone into military research at that time. Now in the post-Soviet period my colleagues, who in those days had no such qualms or reservations, are favored over me in getting grants from foreign foundations. Why should they be preferred to me? Am I supposed to lie and say that my research is actually close to military concerns so that I can get grants? I see that some of my colleagues are doing that, but I am once again being punished for my integrity while they are once more being rewarded for their willingness to do whatever is necessary to get money."[37]

Another indirect influence of foundations on Russian research concerns increasing specificity and justification of research topics, as mentioned in chapter 6. Even in basic research in which the possible topics of study are quite broad, the new specificity of grant-driven research has meant fewer open-ended theoretical investigations. Russian applicants learned that peer review committees are not usually enthusiastic about general statements that one is going to "do theoretical research on condensed matter" or "explore alternative algebras." They prefer promises of concrete products even in fundamental research (specificity of the research on condensed matter based on recent achievements by the same working group; similar further elaboration of a specific mathematical direction, such as Lie algebra, flowing out of recent developments fostered by the applicant).

The old Soviet top-down support of institutions as a whole (block financing) gave great leeway to theoretical investigators within them to go in almost any direction they wished so long as the expenses were not too great. The new grant system brought in from abroad required much more detailed promissory statements in order to win out in competition with others. Researching the effects of such obligatory specificity may not be possible, but one suspects that although most of the effects are positive (greater accountability, more rigorous evaluation), in a few exceptional cases the effects may be negative. In the period 2002–2006 a mathematician in St. Petersburg, Grigorii Perelman, claimed to have solved one of the most important problems in topology, the Poincaré Conjecture. His claims are currently being tested and seem to be valid.

Perelman was awarded the highest prize in mathematics in 2006, the Fields Medal; this includes a financial award, which he turned down because he wished to remain independent. Would Grigorii Perelman, when he was doing this work in the 1990s, have been able to get the support from peer review committees of a foreign foundation in Russia for his claim that he would solve the Poincaré Conjecture? Many such claims had been made before, all, in the final analysis, false. Why would a peer review committee in the nineties have thought this case was different? Perelman was successful in getting several fellowships at universities in the United States (including the University of California at Berkeley), yet in Russia, to the best of our knowledge, he survived only on the old system of block funding of his institute, the Steklov Institute of the Russian Academy of Sciences, miserable though that funding was.

Russian Workers in Foreign Foundations

An under-researched phenomenon, but one increasingly evident, is that the offices of foreign organizations in Russia began to adopt principles and methods of work characteristic of Russian organizations.[38] A process of re-adaptation was occurring, the assuming by foreign structures of Russian management practices. On the one hand, the appearance of foreign foundations brought Russia the procedures of peer review, the concepts of conflict of interest and confidentiality, and other new concepts. On the other hand, in recent years several foreign foundations working for a long time in Russia began to use such methods of work as lobbying "their" grant recipients, changing the rules of choice of grant applicants in competitions, and also breaking the rules of confidentiality. Comments on the formation of a circle of perpetual grant recipients around different foundations increasingly turn up in surveys of scientists and in research on the activities of foundations. An alleviation of this problem might be found in more frequent rotation of experts and also in the collective making of decisions.

In competitions requiring the presentation of letters of recommendation in two languages, it is now accepted in several programs that the applicants themselves can translate such letters into English (or into Russian), a breach of normal rules of confidentiality. Such a practice means, of course, that the applicant knows the contents of the letters of recommendation submitted with his or her proposal, greatly diminishing the value and frankness of those letters. Since in Russia the practice is widespread for letters of recommendation to be written by the applicants themselves, the acceptance by some foreign foundations of this practice can be seen as the absorption of Russian practices. A result is that in various surveys Russian scientists are skeptical about the value of the selection procedures. Today in Russia the most widely held opinion is that the grant system is important and necessary, but that the mechanisms of its practical implementation in Russia—in both foreign and Russian foundations—is not optimal.

The Appearance of Russian Private Philanthropy

Still another unplanned and at the same time positive effect of the work of foreign organizations and foundations is the appearance in Russia of private philanthropy in science. We can say that Russian private philanthropy began in approximately 1999. Its appearance was, of course, not due to the model of foreign foundations alone, but their influence nonetheless can not be denied.

At present in Russia there are several private foundations active in the area of education and science (see table 8.4). However, basic science is currently supported only by the Dynasty Foundation, which gives grants for research in theoretical physics.

Table 8.4. Russian private philanthropy

Name of foundation	Year of establishment; founder	Areas of support	Size and types of grants
Vladimir Potanin Foundation	1999; V. Potanin, President of "INTERROS" holding company	Support of students and young teachers doing research	1,260 stipends of 1,500 rubles a month. Individual grants to teachers of $1,200 on a one-time basis (26 grants in 2003).
Foundation for the Assistance of Domestic Science	2000; presidium of the Russian Academy of Sciences, and the companies Sibneft and Russian Aluminum	Yearly grants to researchers in the Russian Academy of Sciences	Grants to 10 members and corresponding members of the Academy, $10,000 a year (before income tax); grants to 100 researchers with degree of "doktor," $5,000; grants to 100 researchers with degree of "kandidat," $3,000.
Dynasty Foundation	2001; former general director of Vympelkom Dmitrii Zimin	Young scientists (up to 35) and students working in theoretical physics	Yearly grants to 10 young scientists of 15,000 rubles a month; grants to 100 students of 2,000 rubles a month for 9 months.
Alferov Foundation	2001; Academician Zhores Alferov, Nobel laureate	Support of higher education in the natural sciences and one prize a year to a young scientist (up to 33) in greater St. Petersburg.	Stipends to students of 1,200 rubles a month for an academic year. The annual prize to a young scientist is 100,000 rubles.

Sources: I. Dezhina and L. Graham, interviews, 2006–2007; press releases of the foundations.

Even though these organizations have obvious differences, they possess a number of similar characteristics. First of all, they chiefly support individuals, not organizations. They are not interested in how the institutions of the applicants are organized or managed, or how effective they are. In other words, their grants are one-time assistance and have almost no institutional consequences. Also, a number of these organizations pursue commercial and professional goals as well as scientific ones. For example, they are interested in the preparation and selection of personnel for their companies or regions of business activity. Lastly, it should be noticed that all of them give only modest levels of assistance—on the order of one or two million dollars a year. Thus, their impact is also still modest.

It is interesting that until now the only example of direct and obvious influence of foreign foundations on native philanthropy is the New Eurasia Foundation, which appeared at the end of 2004 as an independent legal Russian entity. The founders of this foundation were three organizations: the U.S. Eurasia foundation, the Dynasty Foundation, and the European Madariaga Foundation. The Eurasia Foundation was active in Russia from 1993 to 2004, and the formation of the New Eurasia Foundation was a form of "strategic exit." The basic tasks of this new foundation are the support and strengthening of civil society in Russia, and also assistance to the integration of the country into the world community, including support of scientific and innovation projects and small business.

In general, the direction in which Russian government policy is developing on private philanthropy is worrisome. As will be discussed below, there is a new law on non-governmental organizations that indicates the desire of the Russian government to exercise greater control over philanthropies. On the government level the idea is being actively promoted of the "social responsibility of business." Behind this innocent phrase is the idea of the "governmentalization of philanthropy," according to which the private sector is not so much stimulated toward philanthropy as forced to make donations.

New Law on NGOs

In 2006 the Russian government adopted a new law on non-governmental organizations (NGOs) that has attracted worldwide attention.[39] In the first drafts of this law, the restrictions on NGOs were so severe that some defenders of them feared that either true NGOs in Russia would be impossible or they would be under complete government control. Those who hoped that Russia would develop into a civil society similar to that found in countries of Western Europe or North America saw the new law as a great threat. However, the draft law was revised in such a way that the greatest restrictions were removed, especially for foreign NGOs.

The final law, as adopted by the Duma and signed by President Putin, was implemented in the summer of 2006, but its ultimate results are still not known. Much depends on how the law, which requires NGOs to be registered with the government, will be applied. Will it be used in a strongly restrictive way, or will it be primarily a system of registration that does not change the situation of NGOs in a fundamental fashion? Many worries about the first possibility have been expressed, especially in the Western press. With respect to science and technology, which are the major themes of this book, the new law may not have as much effect as it does on more political topics such as human rights, freedom of speech, and the effort to build a civil society. In any modern democratic society, however, all these elements, including science and technology, are interconnected. Observers in many countries will be watching this situation in the future.

The degree to which foreign foundations have had an impact on Russian science cannot be meaningfully evaluated without considering the different goals those foundations have pursued: philanthropy, partnership, commercialization, security, and reform.

In philanthropy, foreign foundations have had significant effects, particularly in the 1990s, when they helped save Russian science at a time of deep crisis. They have improved the material situation of many Russian scientists and scientific institutions and helped renew scientific equipment. Their efforts to improve geographical distribution of research centers, to help young people, to prevent brain drain, and to foster women scientists have been less clearly successful, although some progress was made. In recent years, philanthropy has become much less important than earlier, replaced more and more by mutually beneficial cooperation. At the same time, foreign organizations served as models for a few emerging Russian philanthropies.

In partnerships, foreign foundations and organizations have helped to create a number of quite successful cooperative efforts between Russian and foreign scientists. The number of publications authored jointly by foreign and Russian scientists has significantly increased. Furthermore, even after the cessation of sponsored partnerships, cooperation among the scientists and institutions involved has often continued. The linkages between Russian and foreign scientists have markedly grown.

Commercialization was not at first a major goal of foreign foundations, but later it became increasingly important, especially after nonproliferation and conversion goals became difficult to pursue. Financing by foreign partners under the programs of the ISTC grew by over 200 times between 1997 and 2005. In the latter year such financing amounted to approximately $207 million U.S., with over 40 percent of the financing in biotechnology and the earth sciences.

The security goal (conversion, nonproliferation) has been supported by very large sums of money, and in the nineties enjoyed moments of success. However, as time went on and the Russian government resumed strongly promoting defense research, organizations like the ISTC could not effectively oppose official government policy. Therefore, the ISTC increasingly turned to the commercialization goal, where its work continues to be significant.

The efforts to achieve reform of Russian scientific and education institutions have yielded mixed, even limited, results. Some progress has been made in developing what is often called "innovation infrastructure" and improving research management practices, but the practical consequences of such efforts are still often unclear or disappointing. The effort to bring teaching and research closer together succeeded in a few universities, but it is still too early to measure lasting effects.

9   Conclusion

During the last fifteen years science in Russia passed through a dramatic period. The collapse of the Soviet Union and the disappearance of the government budget for science threatened it with extinction. Foreign organizations and foundations came to the rescue with the largest science assistance program in history. Hundreds of millions of dollars of foreign money were put into Russian science. Some of the leading Russian research institutes received, for a while, as much as 50 percent of their funds from foreign sources.

But the story of the last fifteen years in Russian science has been much larger than merely a financial crisis. The incredible disruption that occurred created a unique opportunity to observe the anatomy of science. It raised questions that few have understood or pursued. One such question is, "What is the relationship between the political and economic order of a society and its science establishment?" The answer that the Russian example gives is: "The Soviet Union developed a science system that was also 'Soviet' in many respects, centralized and authoritarian." Another relevant question is, "How well does a system of administering and financing science that was developed by an authoritarian government with a centralized economy work in a democratic society with a market economy?" Even though Russia has not completed the transition to democracy and a market economy, already the answer to this second question is clear: "Not very well." Adjusting to the political and economic system of the new Russia was far more traumatic for Russian scientists than any of them had dreamed. Russian science administrators have faced an almost overwhelmingly daunting problem in recent years: At the same time their budgets were disappearing, and many of their best scientists were leaving, they were challenged to devise a new administrative, financial, and legal system in which their remaining scientists could function.

Brain drain has been a major problem in post-Soviet science. Although the problem is not as great now as a few years ago, it is still serious, especially among young people. Quite a few science students in Russia are industriously studying foreign languages, especially English, and many of them are hoping to work abroad at the post-doctoral level. Some will stay abroad if given an opportunity. But recent experience in Russia shows that brain drain can be greatly reduced if proper steps are taken. Those steps are improvement of salaries and equipment; availability of foreign contacts and travel; and career opportunities in Russia. A few research institutions in Russia have taken these steps and become success stories in stemming or reducing brain drain. Many Russian scientists would prefer to live and work in their home country but do not wish to do so at the expense of their families and careers. Brain drain, therefore, is not, from the

Russian standpoint, an unsolvable problem. A stronger economy and improvements in the funding of science can go a long way toward easing the situation.

Sometimes crises are not totally negative in their effects; moreover, they can also present opportunities for change. For example, the financial crisis in Russia in 1998, when the value of the ruble crashed, actually led to the recovery of the Russian economy, which is now growing strongly. However, the crisis that struck Russian science after the fall of the Soviet Union has led to few positive developments in science. There is an apparent paradox here in the fact that crisis caused improvements in Russian economic policy, but did not lead to many positive changes in Russian science policy. (The most notable exceptions were the creation of the RFBR, a government science foundation that is still, unfortunately, underfunded, and the acceptance of peer review on the microlevel, also inadequately supported.) The reasons behind this difference are, as we discuss elsewhere, the conviction of many Russian scientists that their traditional science policy was vastly superior to the traditional Russian economic policy, and also the inertia of Russian scientific institutions. Furthermore, the privatization of economic resources gave industrial officials a large material incentive to tear down the old system; the same incentive was largely missing from the scientific realm.

Nonetheless, there have been many efforts, both foreign and Russian, to help and reform Russian science. How well have they worked, and what have been their effects? Answering these questions adequately is very complicated, since there have been both achievements and disappointments. Perhaps we should start by discussing a few "big" consequences of the reform and assistance efforts, and then proceed to the more numerous "smaller" ones that are, despite their secondary nature, still significant.

The most important consequence of all these efforts is a positive one and fairly simple: Russian science has survived. The Russian science establishment today is much smaller than it was in Soviet times, but it is still not only alive, but quite capable in some areas. The Russian economy is growing, and as a result the government budget for science is increasing; the period of intense foreign aid to Russian science is ending. From now on foreign contacts will be based largely on cooperation and mutual benefit, not philanthropy or assistance.

A second major consequence of foreign influence has been a change in the mentality of Russian scientists about how their work should be supported. During the Soviet period the grant system and peer review so common in the science systems of other nations were unknown in Russia. Scientists depended on the budgets of their research institutions, which were in turn supplied by the central government. Scientists were not expected to submit proposals to outside agencies seeking additional funds.

Today the competitive peer review system is taken for granted by most Russian scientists, and it improves the quality of research. In fact, if Russian scientists have complaints about these new methods of funding science, they are usually about the inadequacy of funds available through them and a perceived lack of fairness in peer review, rather than a criticism of the systems themselves.

Most Russian scientists would like to see the grant system and peer review expand in Russian science even though some of their results have been uncomfortable, forcing many scientists out of science in a Darwinian competition.

Despite the success of the new funding system in changing the mentality of Russian scientists, its appropriate size and influence are still questioned, particularly by administrators who want greater control. On the other hand, reformers are currently calling for liberalization in Russian science—better laws on intellectual property, management based on R&D results rather than administrative authority, budgetary reform, creation of more horizontal linkages among research institutes, universities, and private enterprises.

Foreign funders of research can still play an important role by assuring that Russian scientists—along with those of other nations—are eligible to submit proposals in international competitions. Western governments are sometimes reluctant to do this, since their public funds are understandably usually dedicated to their own citizens and national needs; private foundations active in science have a special role here, providing grants for research to scientists from many nations, as foundations like the Rockefeller Foundation, the Carnegie Corporation of New York, the International Science Foundation, the John D. and Catherine T. MacArthur Foundation, the Wellcome Trust, the Howard Hughes Medical Institute, and other organizations have done in the past. It will be a great mistake if such organizations cease their funding of Russian scientists just because the economic situation in Russia has improved. The autonomy of Russian scientists and the strength of "bottom-up" activity are still at stake. Furthermore, funding Russian scientists is often in foreign interests, since good research often comes from the exchange of ideas. Also, research can usually be co-financed in Russia and therefore is less expensive than doing it in Western Europe or North America alone.

At the same time that these changes have occurred, much of the old Soviet system is still with us. The "three pyramids" of Soviet science—the university system, the Academy of Sciences, and the industrial research institutes—still exist. Changes have been attempted in all three pyramids, but most of them are incomplete. The attempt immediately after the fall of the Soviet Union to make basic changes to the Academy system failed and is only now restarting. Russia still has one of the few systems of fundamental research in the world in which the bestowing of the highest honors in science (election to the most prestigious honorary society) is fused with the administration of research (by the members elected to the honorary society). The university system has also not changed very much. A strong system of research universities has not been created in Russia, despite efforts to do so and a few notable achievements, such as the BRHE program. It is doubtful that such a system can be created as long as the Academy dominates fundamental research.

Perhaps the industrial research pyramid has changed the most, since quite a few research institutions in this pyramid have been privatized, along with the industrial establishment itself. But instead of this resulting in an innovative industrial system with strong research establishments, Russian industry continues to be weak in high technology. The recovery of the Russian economy is the result primarily of extractive industry, particularly oil and gas, and not because it has become a strong competitor in the world economy with a variety of desirable technological products. The old industrial research establishment has been weakened without a new one being strengthened. Exceptions are in the defense area and in the oil and gas industries that compete in international markets.

The greatest disappointments have been in the effort to create a commercial culture in Russia in which innovations are quickly utilized in the economy. The infrastructure for such a commercial culture—financial organizations, innovation centers, personnel retraining centers, patent and legal frameworks, and consulting institutions—are still very inadequately developed. Resolving this problem is probably Russia's greatest challenge, since it cannot rely indefinitely on extraction of natural riches for its prosperity. Somehow it has to find a better way of bringing its impressive strengths in fields such as mathematics, physics, chemistry, and biology to bear on its economy. In these fundamental sciences Russia has greater strengths than many other petroleum-producing countries, such as some of the Middle Eastern nations, but it has so far not found a way to take adequate economic advantage of these strengths.

When we look at the reforms that were attempted, we see that those more intrinsically involving science itself, such as the creation of science foundations and TTOs in universities, are more easily enacted than those related to the economy as a whole, such as venture funds and the promulgation of economic regulations favorable for innovation.

A worry that now haunts a number of thoughtful scientists in Russia is this: Will the effort of the government to help the economy by turning Russian science toward more practical, commercial goals have the effect of blunting the edge of Russian science in the areas where it has traditionally excelled, without creating in their place any significant new strengths? Russian science has always been best in the fundamental disciplines, especially mathematics and theoretical physics. What will be the consequences of the decline in funding and prestige of these subjects, the departure abroad of some their best representatives, and the emphasis of the government on creating a National Innovation System promoting such practical topics as information technology and bioengineering? Will Russian fundamental science decline without Russian high technology becoming genuinely competitive? In the old areas Russia was recognized as one of the best in the world; in the new areas it faces a host of competitors currently doing a better job: the United States, Europe, Japan, China, India, and others. Succeeding internationally in high technology is as much an economic challenge as it is a technical one, and Russia has traditionally not done well at such competitive economic challenges. The prospects for Russia overtaking its rivals in high technology do not look good.

Current trends among foreign foundations active, or previously active, in the former Soviet Union seem to strengthen this danger. The three largest foundations promoting science in the former Soviet Union during the last fifteen years were the ISF, INTAS, and the ISTC, which together invested there well over a billion dollars. The ISF is long gone, and INTAS stopped funding new projects in January 2007. Only the ISTC of the big trio remains. The ISF and INTAS tended to invest in fundamental science; the ISTC traditionally emphasizes applied science (there are many exceptions to this generalization, but in broad terms it is valid). The departure of INTAS will be particularly damaging to the science programs of the non-Russian countries of the former Soviet Union, now independent. In Russia the funding of science is increasing as the economy improves, but the economies of some of the other republics are not improving as rapidly.

In Russia itself much will depend on the ability of the scientific community and the government to create new strengths while holding on to what is left of the old ones. The awarding of the Fields and Clay awards in mathematics to Grigorii Perelman of St. Petersburg in 2006 for his solving of the Poincaré conjecture was a recent triumph of the old spirit of mathematics in Russia: a feat of intellectual genius unrelated to any clear commercial application. The effort of several prominent mathematicians in Russia, such as V. Arnol'd, to preserve Russian mathematical schools by teaching mathematics seminars both in Russia and in other countries (in Arnol'd's case in France and Italy) stems from their desire to benefit from salaries abroad and yet maintain a connection with Russia and young mathematicians there. A dramatic effort is being made to keep the old strengths, but we do not at this moment know how successful it will be.

Some of the problems that afflict Russian science and technology policy are specific to Russia, and some of them are generic in many industrialized countries. Specific problems in Russia are connected to the great size of the country, the largest in geographic terms in the world, and the legacy of the Soviet Union, which still affects the mentality of many Russian scientists and science administrators. General problems affecting many countries, including Russia, are minimizing market and government failures. Russia shares with other countries what is sometimes called the European Paradox, the simultaneous coexistence of strong science and low commercialization activity. How does a country take commercial advantage of its scientific achievements? This is a question being studied throughout Europe.

Paradoxically, one of the reasons reforming science and education in Russia has been so difficult is that they were relatively strong in the Soviet period. At the time of the collapse of the Soviet Union there was almost universal agreement among Soviet citizens and their leaders that the Soviet political and economic systems had been failures. Science and education were, however, often praised as "the best things the Soviet Union did." Therefore science and education were tied to national pride in a way that politics and economics were not. This pride in past achievements was an obstacle in understanding that the old Soviet systems of research and education, meritorious though they may have been in the past, simply would not work in the present globalized, economically competitive age. The best Soviet science was divorced to a large degree from the economy and from the education of youth. It was organized and funded in ways that best practices in other countries increasingly demonstrated were outdated. But the pride of Soviet scientists and educators often prevented them from seeing the obvious. When strong reform efforts were made after the fall of the Soviet Union, the old scientific establishment often strenuously objected, bewailing "what we are about to lose—the glory of Soviet education and science."[1] The long delays in the reform of the Academy of Sciences and in bringing teaching and research closer together are at least partially attributable to these attitudes.

On the other hand, the question of the degree to which foreign models can be transferred to another country is a legitimate one. Often when foreign advisors talk to Russian scientists and administrators about such things as integrating research and education, both sides agree on the desirability of such a goal, but each side understands it in a different way. Russian scientists often take integration to mean providing ways for research scientists from the Academy of Sciences to teach some courses in universities, at the same time looking for the best students who can be transferred, at the graduate stage, to the Academy system. In this understanding the universities are feeders to the Academy. Integration, according to this view, means that there would be many personal connections between universities and academy institutes, but most teaching would still go on in the universities and most research would still take place in the Academy. Western, especially U.S., advisors who speak of integration of research and education usually mean that both activities should go on *in the same place,* that the best practice for teaching and research is when they exist in symbiotic relationship in the same institution. Teachers should be researchers as well as teachers, and the people they teach should be both students and researchers at the same time. This different understanding is still often present, although a few Western programs, such as the Basic Research and Higher Education Program, have made some progress in overcoming it in a few Russian universities.

Since the possibility of creating a strong and large system of research universities of the U.S. type in Russia is small, perhaps the best hope for Russia in this area is the creation of something entirely different: a unique system in which the universities and the academy institutes have many mutual ties and much temporary exchange of personnel but in which each retains its primary responsibility for, respectively, education and research. Surely the U.S. system is not the only system in which good advanced research and education can occur. France and Germany, with their research establishments, respectively, of the CNRS and the Max-Planck-Gesellschaft, may have more appropriate experience for Russia in this respect than the United States does, especially since both France and Germany are currently trying to reform their own systems in a way that will bring teaching and research closer together and make their science establishments more competitive in the world. Germany's recent decision to strengthen selected research universities seems particularly pertinent for Russia.

What is the "bottom line" evaluation of the efforts during recent decades to reform Russian science, culture, and politics? Not surprisingly, such an evaluation has to be mixed. Without any question Russia today is in many respects a dramatically different country than it was twenty years ago. There is no formal system of censorship, and although most forms of mass media are now government-controlled, there are still newspapers and radio stations that are free and regularly discuss very controversial political questions. The economy is now definitely a mixed one, with a great variety of small private enterprises, especially in retail and service trades. In science, the area on which this book has concentrated, new methods of funding and administering science have arisen, competing with the traditional ones. Perhaps most important of all, Russian scientists are much more integrated into the world scientific community. They are free to travel, if they can find the necessary funds, and their links with foreign laboratories and foreign colleagues are multitudinous. Russian science today is more a part of world science than it has ever been before.[2]

Thus, for all the disappointments, inadequacies, and remaining problems facing Russian science, significant achievements have been made. The drama of how one of the world's largest scientific communities copes with a strikingly different political, social, and economic context than the one in which its major formation occurred is still with us.

Notes

Introduction

1. In 1993 the prominent Russian physicist Aleksei Abrikosov wrote, "I am certain that helping science in Russia is hopeless. . . . Today there is only one way to preserve Russian science: help all talented scientists to leave Russia as soon as possible, and forget about the rest." "Akademik Abrikosov nakhodit pliusy v 'utechke mozgov' na zapad" (Academician Abrikosov finds pluses in the 'brain drain' to the West) and "Ia nikogda ne vernus' v Rossiiu" ("I will never return to Russia"), *Izvestiia,* May 5, 1993. English translation of Russian newspaper articles can often be found in the *Current Digest of the Post-Soviet Press.*

2. Gloria Duffy, chair of the board, Civilian Research and Development Foundation (CRDF), as quoted by Richard Stone in "Cautious Optimism, But Progress is Slow," *Science* 294 (November 2, 2001): 974.

3. Examples are: Harley D. Balzer, *Soviet Science on the Edge of Reform* (Boulder, Colo.: Westview Press, 1989); Glenn E. Schweitzer, *Swords into Market Shares* (Washington, D.C.: Joseph Henry Press, 2000); and Gerson Sher, "Why Should We Care About Russian Science?" *Science* 289 (July 21, 2000).

1. Science at the End of the Soviet Period

This chapter draws on material in several earlier publications by one of the authors: Loren R. Graham, "Big Science in the Last Years of the Big Soviet Union," *Osiris* 7 (2nd series, 1992): 49–71; *What Have We Learned About Science and Technology from the Russian Experience?* (Stanford: Stanford University Press, 1998); *Science in Russia and the Soviet Union: A Short History* (Cambridge: Cambridge University Press, 1993).

1. Dmitry I. Piskunov, "Soviet Fundamental Science: State, Problems, and Perspectives of Development," unpublished manuscript, in Russian (Moscow: Analytical Center for Problems of Socio-Economy and Science-Technology Development of the Academy of Sciences of the USSR, 1991), 2.

2. Louvan Nolting and Murray Feshbach, "R&D Employment in the USSR," *Science* 207 (February 1, 1980): 493–503.

3. Loren R. Graham, "Big Science in the Last Years of the Big Soviet Union," *Osiris* 7 (2nd series, 1992): 56.

4. See Mark Adams, "Research and the Russian University," in *The Academic Research Enterprise Within the Industrialized Nations: Comparative Perspectives: Report of a Symposium* (Washington, D.C.: National Academy Press, 1990), 51–65. There were, of course, exceptions to the above generalizations. Moscow and Leningrad universities, for example, always had some outstanding researchers,

especially in fields like mathematics where talented young students often play significant roles.

5. Alexander Vucinich, *Empire of Knowledge: The Academy of Sciences of the USSR (1917–1970)* (Berkeley/Los Angeles: University of California Press, 1984).

6. Boris Saltykov, "The Reform of Russian Science," *Nature* 388 (July 3, 1997): 16.

7. Piskunov, "Soviet Fundamental Science," 2.

8. A graphic, at moments lurid, description of such corruption is in Mark Popovskii, *Manipulated Science: The Crisis of Science and Scientists in the Soviet Union Today,* Paul S. Falla, trans. (Garden City, N.Y.: Doubleday, 1979).

9. The chairman of the committee conducting the study was Dr. Carl Kaysen, Director of the Institute of Advanced Study in Princeton. The final published report is *Review of U.S.-U.S.S.R. Interacademy Exchanges and Relations* (Washington, D.C.: National Academy of Sciences, 1977).

10. Péteri's graph is reproduced in Loren R. Graham, *What Have We Learned About Science and Technology from the Russian Experience?* (Stanford: Stanford University Press, 1998), 158.

11. Nolting and Feshbach, "R&D Employment in the USSR," 493–503.

12. Thane Gustafson, "Why Doesn't Soviet Science Do Better Than It Does?" in Linda Lubrano and Susan Solomon, eds., *Social Context of Soviet Science* (Boulder, Colo.: Westview Press, 1980); and F. M. Borodkin et al., eds., *Postizhenie: Sotsiologiia, sotsial'naia politika, ekonomicheskaia reforma* (*Comprehension: Sociology, social policy, economic reform*) (Moscow: Progress, 1989), 140–168.

13. Nikolai Bukharin, "Theory and Practice from the Standpoint of Dialectical Materialism," *Science at the Cross Roads* (London: Kniga, 1931), 23.

14. J. D. Bernal, *The Social Function of Science* (London: Routledge, 1939); J. G. Crowther, *Soviet Science* (London: K. Paul, Trench, Trubner, 1936).

15. Bernal, *The Social Function of Science,* and Chris Freeman, "The Social Function of Science," in Brenda Swann and Francis Aprahamian, eds., *J. D. Bernal: A Life in Science and Politics* (London and New York: Verso, 1999), 125.

16. Henry E. Sigerist, *Socialized Medicine in the Soviet Union* (New York: Norton, 1937), 308.

17. Quoted in Gary Werskey, "The Radical Critique of Capitalist Science: A History in Three Movements," paper prepared for Princeton conference "Marxism and Science," Princeton History of Science Workshop (March 31–April 1, 2006), 29.

18. George. W. Corner, *A History of the Rockefeller Institute: 1901–1953, Origins and Growth* (New York: The Rockefeller Institute Press, 1964), 39.

19. Ibid., 9.

20. S. F. Ol'denburg, "Vpechatleniia o nauchnoi zhizni v Germanii, Frantsii i Anglii," (Impressions of scientific life in Germany, France, and England) *Nauchnyi Rabotnik* (February 1927): 89.

21. George W. Corner, *A History of the Rockefeller Institute,* 541; and conversations

with Alexander Bearn, former professor and trustee of the Rockefeller Institute and Rockefeller University, 2003–2005.

22. Conversations with Alexander Bearn, 2003–2005.

23. Toni Scarpa, "Peer Review at NIH," *Science* 311 (January 6, 2006): 41.

24. Richard Feynman, *Surely You're Joking, Mr. Feynman! Adventures of a Curious Character* (New York: Bantam Books, 1986).

25. Roald Hoffmann, "Research Strategy: Teach," *American Scientist* 84 (January-February, 1996): 20. See also Hoffmann's "University Research and Teaching: An Enriching and Inseparable Combination," *Boston Sunday Globe,* November 5, 1989.

26. See the forthcoming history of Lusitania: Loren Graham and Jean-Michel Kantor, *Naming God, Naming Infinities* (Cambridge, Mass.: Harvard University Press, expected publication 2009).

27. See, for example, the analysis and recommendations given in *Forschungsförderung in Deutschland: Bericht der internationalen Kommission zur Systemevaluation der Deutschen Forschungsgemeinschaft und der Max-Planck-Gesellschaft* (Hannover: Neue Medien, 1999), especially pp. 15–20.

28. See Aisha Labi, "Germany Awards 'Elite' Status and $26 Million to 3 Universities," *The Chronicle of Higher Education,* October 27, 2006, A42.

29. Examples of outspoken views from this group in 1989–1991 were Maksim Frank-Kamenetskii, L. Tausen, and Boris Raushenbakh. Graham/Sloan Archive B2, p. 317; B2, p. 323; B2, p. 325. The Alfred Sloan Foundation financed the collection of materials about the relationship of Russian science and the breakup of the Soviet Union with a grant to Loren Graham. The resulting archive is now on deposit in two different places: the library of the Institute of the History of Science and Technology of the Russian Academy of Sciences in Moscow, and Harvard University (catalogued under "Graham/Sloan Archive" and also cited here in the same way). Most of the articles are in Russian, but translated titles are presented here.

30. V. D. Kazakov, "Science and Measures of Truth," (Feb. 5, 1989), Graham/Sloan Archive, B2, p. 318.

31. Perhaps the most outspoken and intelligent exponent of these views was Aleksei Zakharov from the Institute of Oceanography of the Academy, the leader of several different unions of scientific researchers. See, e.g., his "Do We Need a Professional Union of the Academy of Sciences?" (Nov. 1, 1990), Graham/Sloan Archive 11.17, B8, p. 194.

32. V. Ginzburg, a full member, noted the resentments of the corresponding members, and favored an elimination of the dual ranks. However, he recognized the difficulty in achieving this change, and recommended as a short-term easing of the problem the granting of voting rights in the General Assembly of the Academy to corresponding members (V. Ginzburg, "Democracy in the Style of the Academy" [April 14, 1990], Graham/Sloan Archive, 7.10, B2, p. 327).

33. The resentment of the engineers was discussed by K. Frolov in "A Russian Academy" (Dec. 29, 1989), Graham/Sloan Archive, 7.10, B2, p. 327. The Academy of Sciences before 1929 did not have engineers among its members

and did not engage in engineering research. The first engineers were elected in that year; between 1929 and the late 1950s their numbers grew until the Department of Technical Sciences (engineering) had more members than the three departments for social sciences and humanities combined. In the sixties, after a great controversy, the engineers were expelled. See Loren R. Graham, "Reorganization of the U.S.S.R. Academy of Sciences," in Peter H. Juviler and Henry W. Morton, eds., *Soviet Policy-Making: Studies of Communism in Transition* (New York: Frederick A. Praeger, 1967), 133–162.

34.　L. Maiboroda represented this view in "Science Without a Ministry of Science" (May 18, 1991), Graham/Sloan Archive, 7.55.

35.　Resolution of the Supreme Soviet of the Russian Republic "On the Rules for the Formation and Organization of the Activity of the Academy of Sciences of the RSFSR" (July 13, 1990) (signed by Yeltsin), Graham/Sloan Archive 2.1.

36.　V. Tishkov, "Is This the Academy or a Fortress?" ("Akademiia ili bastion?" *Sovetskaia Kul'tura* [April 28, 1990]), Graham/Sloan Archive 7.20, B2, p. 339.

37.　Gorbachev's decree was "On the Status of the Academy of Sciences of the USSR" (Aug. 23, 1990), Graham/Sloan Archive 2.2, B2, p. 116. The Supreme Soviet of the RSFSR canceled the part of Gorbachev's decree that applied to property in Russia: Resolution of the Supreme Soviet of the RSFSR, "On Canceling the Validity of Article 2 of the Order of the President of the USSR" (Sept. 21, 1990), Graham/Sloan Archive 2.3. Yeltsin's cancellation of Gorbachev's decree in entirety is discussed in Andrey Allakhverdov, "Russian Scientists Gain Legal Rights . . . ," *Science* 273 (July 26, 1996): 424.

38.　Iu. Danilin, "Academic Horrors" (October 22, 1990), Graham/Sloan Archive 7.36, B2, p. 508; and G. Tolstikov, V. Kazakov, Iu. Manakov, V. Napalkov, and O. Chupakhin, "Their Own Academy is Closer to Business?" (Feb. 27, 1991), Graham/Sloan Archive 7.49, B2, p. 530.

39.　See footnote 29 of this chapter; as described there, the archive is now located both at Harvard University (catalogued in the library system under "Graham/Sloan Archive") and at the Institute of the History of Science and Technology of the Russian Academy of Sciences in Moscow.

40.　V. Ginzburg, "Democracy in the Style of the Academy" (April 14, 1990), Graham/Sloan Archive 7.17, B2, p. 335.

41.　V. Tishkov "Is This the Academy or a Fortress?" 339.

42.　"O Give Us Freedom!" Graham/Sloan Archive, B2, p. 511, 7.39; ("O daite nam svobodu!"), *Poisk* no. 46 (November 16–22, 1990), 5.

43.　A. Burshtein, "Why does our science nurture academicians and foreign science nurture Nobel Prize winners?" ("Pochemu nasha nauka vskarmlivaiet akademikov, a zarubezhnaia—nobelevskikh laureatov?"), Graham/Sloan Archive, B2, p. 510, 7.38.

44.　Quoted by Aleksei Zakharov, "The Democratic Opposition in the Process of the Creation of the Russian Academy of Sciences," unpublished paper, MIT workshop (April 23–24, 1993), from *Nauka v Sibiri* no. 2 (1991).

45.　Zakharov, "Democratic Opposition," 38.

46.　Ibid., 32.

2. Breakup of the Soviet Union and Crisis in Russian Science

1. Center for Science Research and Statistics (CSRS), *Nauka v SSSR* (1991), 256.

2. Russian Economic Reform: A World Bank Country Study (1992), 18.

3. B. Saltykov, "The Reform of Russian Science," *Nature* 388 (July 3, 1997): 16.

4. *Istochnik dannykh: Nauka v Rossiiskoi Federatsii* (*Source of statistics: Science in the Russian Federation*), Goskomstat RF (Moscow, 1995), 75. According to former Minister of Science M. Kirpichnikov, the equipment base of science continued to worsen throughout the 1990s: see *Poisk* no. 9 (March 3, 2000): 2.

5. Alferov gave a graph to Loren Graham illustrating the budget of his institute from 1985 to 1994. It is reproduced in Loren R. Graham, *What Have We Learned About Science and Technology from the Russian Experience?* (Stanford: Stanford University Press, 1998), 66.

6. *Russian Science and Technology at a Glance: 1999* (Moscow: Data Book/CSRS, 1999), 37.

7. *Russian Science and Technology at a Glance: 2000,* 40.

8. *Russian Science and Technology at a Glance: 2003,* 69.

9. E. Nekipelova, L. Gokhberg, and L. Mindeli, "Emigratsiia uchenykh: problemy i real'nye otsenki" ("The emigration of scientists: Problems and realistic evaluations") in *Migratsiia spetsialistov Rossii: prichiny, posledstviia, otsenki* (Moscow: Institute of Economic Forecasting, Russian Academy of Sciences [INPRAN] and Rand Corporation, 1994), 10; also O. Ikonnikov, *Emigratsiia nauchnykh kadrov iz Rossii: segodnia i zavtra* (*The emigration of scientists from Russia: Today and tomorrow*) (Moscow: Kompas, 1993), 90, 98.

10. E. Krasinets, *Mezhdunarodnaia migratsiia naseleniia Rossii v usloviiakh perekhoda k rynku* (*The international migration of the population in Russia in market-transition conditions*), Population Institute, Russian Academy of Sciences, 1997.

11. I. Ushakov and I. Malakha, *Utechka umov: masshtaby, prichiny, posledstviia* (*Brain drain: Dimensions, causes, consequences*) (Moscow: Editorial URSS, 1999), 71.

12. A. Iurevich and I. Tsapenko, *Nuzhny li Rossii uchenye?* (*Does Russia need scientists?*) (Moscow: Editorial URSS, 2001), 76.

13. *Poisk* no. 47 (November 26, 1999): 1.

14. Institute of International Education (IIE), *Open Doors—1997/1998: Report on International Education Exchange* (New York: IIE, 1999), 15.

15. Center for Sociological Research, "Deiatel'nost sotrudnikov nauchnykh podrazdelenii vuzov" ("The work of university science department employees"), Ministry of Education of the Russian Federation (December 2000), 24.

16. E. Mirskaia, "Rossiiskaia akademicheskaia nauka v zerkale sotsiologii: Empiricheskie issledovaniia 1994–1999 godov" ("The Russian Academy of Sciences in the mirror of sociology: Empirical research 1994–1999"), *NG-Nauka* (May 24, 2000). Young researchers cited both the decrease in the number of

research and development vacancies abroad and their disappointment with Western research approaches as reasons for the decrease in permanent emigration.

17. This survey was conducted among 500 Moscow researchers in all fields of science and in all types of research organizations. The situation outside Moscow was probably worse; see E. Krasinets and E. Tiuriukanova, "Intellektual'naia migratsiia" ("Intellectual migration"), *Ekonomist* no. 3 (1999): 75.

18. *Nauka Rossii v tsifrakh—2003; Statisticheskii sbornik* (Moscow: TsISN, 2003), 63; *Science and Engineering Indicators 2004,* National Science Board, 2004, Arlington, VA: NSF, volume 2, pp. A3–A31.

19. *Nauka Rossii v tsifrakh—1996: Kratkii statisticheskii sbornik,* (Moscow: TsISN, 1996), 19.

20. Scott Smallwood, "Doctor Dropout," *The Chronicle of Higher Education,* January 16, 2004.

21. More detailed results of the surveys are in I. Dezhina, "Molodye kadry v rossiiskoi nauke: kak ikh sokhranit'?" ("Young people in Russian science: How to preserve them?"), *Naukovedenie* no. 2 (2003): 127–138, and I. Dezhina, "Molodezh' v nauke," ("Youth in science"), *Sotsiologicheskii zhurnal* no. 1 (2003): 71–87.

22. *Russian Science and Technology at a Glance: 2000,* p. 31.

23. Denis Prokopenko, "Kas'ianov vstretilsia so studentami" ("Kasianov met with students"), *Nezavisimaia gazeta,* April 19, 2001, 2.

24. *Poisk* no. 14 (April 13, 2001), 7.

25. "Annual Message of the President of the Russian Federation to the Federal Assembly," April 3, 2001.

3. Major Directions of Reform in Russian Science

1. B. Saltykov, "Aktual'nye voprosy nauchno-tekhnicheskoi politiki" ("Current issues in S&T policy"), *Naukovedenie* (Nov. 1, 2002): 57.

2. Calculated from statistics in *Nauka v Rossii v 1996 godu* (Moscow: Goskomstat RF, 1996), 18; *Nauka Rossii v tsifrakh—1999* (Moscow: TsISN, 1999), 37.

3. *Nauka Rossii v tsifrakh—2001: statisticheskii sbornik* (Moscow: TsISN, 2001), 45.

4. Irina Dezhina, "Reform of Public Sector Science: A New Government Concept" (in Russian), *Ekonomiko-politicheskaia situatsiia v Rossii* (Moscow: Institute for the Economy in Transition, 2004), 29–33.

5. Andrey Allakhverdov and Vladimir Pokrovsky, "Kremlin Brings Russian Academy of Sciences to Heel," *Science* 314 (November 10, 2006): 917.

6. "Rossiiskaia akademiia nauk lishaetsia nezavisimosti" ("The Russian Academy of Sciences loses independence"), *Kommersant,* December 6, 2006.

7. Anton Arnol'dovich Trofimov, "Patronazhnaia natsionalizatsiia" ("Patronage nationalization"), *Nezavisimaia gazeta,* December 7, 2006.

8. Nadezhda Volchkova, "So znakom plius: izmeneniia v zakone o nauke gosakademiiam ne povrediat" ("A plus: The changes in the law on science do not harm the state academies"), *Poisk* no. 50 (December 15, 2006), 3, 6.

9.   Ibid., 6.

10.   Gorbachev's decree was "On the Status of the Academy of Sciences of the USSR" (Aug. 23, 1990), Graham/Sloan Archive 2.2, B2, p. 116. The Supreme Soviet of the RSFSR canceled the part of Gorbachev's decree that applied to property in Russia: resolution of the Supreme Soviet of the RSFSR, "On Canceling the Validity of Article 2 of the Order of the President of the USSR" (Sept. 21, 1990), Graham/Sloan Archive 2.3. Yeltsin's cancellation of Gorbachev's decree in its entirety is discussed in Andrey Allakhverdov, "Russian Scientists Gain Legal Rights . . . ," *Science* 273 (July 26, 1996): 424.

11.   On November 19, 2007, the Russian government confirmed a new charter for the Russian Academy of Sciences that included the provisions discussed in the preceding section. According to that charter, the president of the Academy is elected by the General Assembly and confirmed by the President of the Russian Federation; the same procedure would be followed for any new charter. The Academy would retain much autonomy, without the sort of "supervisory committee" suggested by some more radical reformers. In sum, the Academy retains many of its privileges, and even has some new ones in the area of education and applied science, as discussed in the above paragraph. *Postanovlenie Pravitel'stva RF "O Rossiiskoi akademii nauk,"* No. 785, November 19, 2007.

12.   Data for 2002. *Nauchnyi potentsial i tekhnicheskii uroven' proizvodstva,* No. 3, RUDN, Moscow, 2005, p. 6.

13.   See I. G. Dezhina, *Mekhanizmy gosudarstvennogo finantsirovaniia nauki v Rossii (Mechanisms of state financing of science in Russia)* (Moscow: Institut ekonomiki perekhodnogo perioda, 2006).

14.   In Russian, Federal'naia Tselevaia Nauchno-Tekhnicheskaia Programma (the federal goal-oriented S&T program) (FTsNTP).

15.   For evaluations of corruption in S&T see, for example, E. Pis'mennaia, "Otkati, togda pokatit" ("Bribe, then things move"), *Russkii Newsweek* no. 57 (July 25–31, 2005): 17–20; E. Morgunova, "Vpered, k ruinam? Gosudarstvo khochet izbavit'sia ot otraslevoi nauki" ("Forward to ruin? The government wants to get rid of industrial science"), *Poisk* no. 25 (June 24, 2005), 4.

16.   E. Nekipelova, L. Ledeneva, "Russkii student na eksport" ("Russian students for export"), *Inostranets* no. 8 (March 11, 2003).

17.   The data were collected in 2003. E. Nekipelova, L. Ledeneva, "Okhota na umy: proigrannyi raund" ("The search for brains: Lost round"), *Poisk* no. 46 (November 14, 2003), p. 6.

18.   *OECD Science, Technology and Industry Outlook,* OECD (Organization for Economic Cooperation and Development), 2004, 79–80.

19.   Caglar Ozden, Maurice Schiff, eds., *International migration, remittances, and the brain drain* (Washington D.C.: World Bank, Palgrave Macmillan, 2005).

20.   David Spurgeon, "Canada Tries to Limit Nanotech Brain Drain," *Nature* 409 (November 30, 2000): 623; Wayne Kundro, "Canadian Universities: Funding of 2000 Slots Sets Off Musical Chairs," *Science* 288 (June 23, 2000): 2112.

21.   Jeff Chu, "How to Plug Europe's Brain Drain," *Time* (January 13, 2004).

22.   David Heenan, *Flight Capital: The Alarming Exodus of America's Best and Brightest* (Mountain View, Calif.: Davies-Black Publications, 2005).

23.   *Prism* 5, no. 14 (1999); Amnesty International News Service No. 012, January 19, 2007.

4. Foundations

1.   Gennady Kochetkov and Igor Nikolaev, "The Russian Science Foundation: An Unfinished History," MIT Workshop, 1994, p. 9. An organizer of these visits on the U.S. side was Harley Balzer of Georgetown University.

2.   Conversations of Loren Graham with DFG officials, Bad Godesbug, February 4, 2004. In subsequent discussion with RFBR officials Irina Dezhina was told that the major model for the RFBR at the time that it was created with the NSF, but that after the RFBR had started its operations its administrators also took into account the experience of some other foreign organizations and foundations, including the DFG.

3.   Frederick Paul Keppel, "Extracts from the Annual Reports of the President of the Carnegie Corporation, 1931–35," in *Philanthropy and Learning with Other Papers,* (New York: Columbia University Press, 1936), 74.

4.   This is a growth in comparison with the initial level. Until 1997 the RFBR received 4 percent of federal budget expenditures for civilian science, and the RFH received 0.5 percent. After 1997 RFBR started to receive 6 percent and RFH 1 percent.

5.   *Novosti RFBR* no. 5 (August 1998), 3.

6.   In 2004 the law was interpreted for a while in a way that would prohibit such joint financing. Now that difficulty seems to be eased.

7.   V. Konov, L. Lialiushko, A. Blinov, "Rossiiskii fond fundametal'nykh issledovanii: 14 let sluzheniia rossiiskoi nauke" ("The RFBR: 14 years of service to Russian science"), *Materialy mezhdunarodnoi konferentsii* (Kiev: Fenix, 2006), 147.

8.   V. I. Bakhmin, *On Foundations in Russia* (Moscow: Logos, 2004), 59.

9.   *Nauka Rossii v tsifrakh—2003,* statistical collection (Moscow: TsISN, 2003), 55.

10.   On average more than 60 percent of the grants of the RFBR and more than 50 percent of the grants of the RFH go to researchers in academic institutes.

11.   See report of Sergei Belanovskii about the Russian Academy of Sciences, "Otsenka sostoianiia Rossiiskoi Akademii Nauk: kratkii otchet" ("An evaluation of RAN: A short report"), www.polit.ru/science/2005/12/15/ran.html (accessed October 2007).

12.   The survey was conducted among more than 300 researchers with the "kandidat" degree and higher. In addition, fifty in-depth interviews were conducted. Also, see I. Dezhina, *Polozhenie zhenshchin issledovatelei v rossiiskoi nauke i rol' fondov* (*The situation of women researchers in Russian science and the role of foundations*) (Moscow: TEIS, 2003).

13.   A. Alakhverdian, I. Dezhina, A. Iurevich, "Foreign sponsors of Russian science:

vampires or Santa Claus?" *Mirovaia ekonomika i mezhdunarodnye otnosheniia* no. 5 (1996), 35–45.

14. E. V. Semenov, *Reality and Day-dreams in Russian Science,* (in Russian) (Moscow: Nauka, 1996), 73.

15. G. S. Batygin, "The invisible boundary: grant support and the reconstruction of scientific society in Russia (remarks of an expert)," *Naukovedenie* no. 4 (2000), 72.

16. The listings are of the following form: "Ivan Ivanov, Moscow State Univeristy." However, the listing does not give the rank or administrative position of the principal investigator, and even the institutional listing may be misleading, since the institution given is the one through which the RFBR finances the project, not necessarily the one in which the principal investigator actually works. It is rather often the case that the PI works in one institute but the financing for project is directed through another, since some organizations charge lower overhead than others and some have a more friendly relationship with the RFBR.

17. I. Dezhina, *The situation of women researchers in Russian science.*

18. Listed in descending order by the number of respondents who cited them.

19. The results summarized in tables 4.4 and 4.5 are based on the survey that Irina Dezhina conducted in 2002–2003 in ten regions of Russia (Moscow, St. Petersburg, Vladivostok, Ekaterinburg, Kazan, Nizhny-Novgorod, Novosibirsk, Saratov, Tomsk, Ufa). Questionnaires were sent to grant recipients of Russian and foreign foundations; 307 responses were received; the rate of return was 55 percent. The age distribution of answers correlated with the overall age structure. The level of qualification of recipients was higher than average—usually these recipients were candidate or doctoral degree holders with positions not lower than "researcher" and "senior researcher."

20. See note 19 for table 4.4.

21. Derek Bok, "Mute Inglorious Wizards," *New York Times* (November 17, 1996).

22. The greatest such controversy involving NSF was undoubtedly that of MACOS ("Man: A Course of Study), peaking in 1975, which involved the social sciences. See Peter Dow, "MACOS: Social Studies in Crisis," *Educational Leadership* 43, no. 1 (October 1979).

23. April 27, 1992.

24. M. Alfimov, V. Minin, A. Libkind, "Strana nauki—RFBR" ("A country of science—the RFBR"), in M. Alfimov and V. Novikov, eds., *Granty RFBR: rezul'taty i analiz* (Moscow: Yanus-K, 2001), 13.

25. Federal Law "Concerning the Introduction of Changes in Articles 24, 158, and 160 of the Budget Code" (No. 158-FZ of December 8, 2003).

26. *Poisk* no. 11 (March 16, 2007), 3.

27. Saltykov and Dezhina have worked out these ideas together. See I. Dezhina and B. Saltykov, "On efficiency in the use of budget funds in Russian science," (in Russian) *Ekonomika i matematicheskie metody* 38, no. 2 (2002): 36–48.

5. Developing a Commercial Culture for Russian Science

1. Data from the Ministry of Education and Science of the Russian Federation.

2. *Science in Russia at a Glance: Statistical Yearbook* (Moscow: CSRS, 2003), 215.

3. *OECD Science, Technology and Industry Outlook,* OECD, 2004, pp. 41–42.

4. Later the Ministry of Industry, Science, and Technology, and still later the Russian Agency for Science and Innovation.

5. Industries cooperate on R&D projects that serve several industries simultaneously. They do this by forming cooperative funds that in the Russian language are called "inter-branch funds" (*mezhotraslevye fondy*), a term that is perhaps translated into English more felicitously as "inter-industrial funds."

6. *Certificate about the system of industrial fund for financing R&D by the Russian Foundation for Technological Development* (Moscow: Minpromnauki RF, 2002).

7. Resolution No. 65 of February 3, 1994.

8. See the SBIR Gateway: http://www.zyn.com/sbir/ (accessed November 2007).

9. See the ANVAR site, www.agence-nationale-recherche.fr (accessed November 2007).

10. I. Bortnik, "10 years of the development of small innovation entrepreneurship in Russia," *Innovation* no. 1 (2004): 5.

11. Irina Dezhina, "Sostoianie sfery issledovanii i razrabotok v 2005 godu" ("The state of R&D in 2005"), *Rossiiskaia ekonomika v 2005 g.: tendentsii i perspektivy,* no. 27, IET, Moscow, 2006.

12. Order No. 362 of March 10, 2000.

13. *Economic Newsletter,* Davis Center for Russian and Eurasian Studies, Harvard University (May 16, 2005), 1.

14. A. Nikkonen, I. Karzhanova, A. Shastitko, eds., *Development of Venture Investments in Russia: the Role of the State* (in Russian), (Moscow: TEIS, 2004), 38.

15. *Rossiiskaia gazeta,* October 8, 2002, 5–6.

16. "Interv'iu nachal'nika otdela Departamenta strategii sotsial'noekonomicheskikh reform Minekonomrazvitiia Rossii A. E. Shadrina, 'Shest' program podderzhki innovatsii'," supplement to the newspaper *Vedomosti* "Forum," December 2005.

17. O. Golichenko, "The innovation system of Russia: a model and the future possibilities of its development" (Moscow: RUDN press, 2003), 181.

18. *Poisk* no. 33–34 (August 25, 2000), 13.

19. Irina Dezhina, "Sostoianie sfery issledovanii" ("The state of R&D in 2005").

20. Data from the Fund for Assistance.

21. Federal law of July 22, 2005, No. 116-FZ "Ob osobykh ekonomicheskikh zonakh v Rossiiskoi Federatsii" ("On special economic zones in the Russian Federation"); and Resolution of the Government of the Russian Federation of September 13, 2005, No. 563 "Ob utverzhdenii Polozheniia o provedenii konkursa po otboru zaiavok na sozdanie osobykh ekonomicheskikh zon" ("On confirming the Statute for competitive selection of special economic zones").

22.   Aleksei Shapovalov (interview of Iurii Zhdanov), "Luchshe komu-to otkazat', chem. vviazat'sia v avantiuru," ("Better to refuse someone than get involved in an adventure"), *Kommersant* (September 15, 2005), 15.

23.   A. Watkins "From knowledge to wealth: Transforming Russian science and technology for a modern knowledge economy," *World Bank Working Paper #2974,* WB Policy Research, 2003.

24.   See Derek Curtis Bok, *Universities in the Marketplace: The Commercialization of Higher Education* (Princeton: Princeton University Press, 2003).

25.   See the entire issue of *Technology Review,* published by the Massachusetts Institute of Technology, in which leading lawyers take opposite points of view on questions of "open source software" and protection of digitalized information. *Technology Review,* June 2005.

26.   See "Conception of the Development of Venture Industry in Russia," in http://www.thinktank.ru/documents.asp?ob_no=463&d_no=480 (accessed in 2007; at time of publication, website was not available).

27.   I. Dezhina, "Sostoianie sfery issledovanii" ("The state of R&D in 2005"), 328–329, citing data from the Federal Agency for Science and Innovations.

28.   The Patent Law of the RF, coming into effect on Oct. 14, 1992, by resolution of the Supreme Soviet of RF of Sept. 23, 1992, No. 3517-1 (henceforth—Patent Law). Law of the RF "Concerning trademarks, service marks and the designation of places of origin of goods," of Sept. 23, 1992 No. 3520-1; Law of the RF "Concerning legal protection for computer programs and databases," of Sept. 23, 1992 No. 3523-1; Law of the RF "Concerning protection of the topology of integrated microsystems," of Sept. 23, 1997; Law of the RF "On competition and restricting of monopolistic activity in goods markets," (in editing of laws of the RF of June 24, 1992 No. 3119-1 of July 15, 1992 No. 3310-1, FZ of May 25, 1995 No. 83-FZ, FZ of May 6, 1998 No. 70-FZ); Law of the RF "concerning author's rights in contiguous rights" of July 9, 1993 No. 5351-1; Law of the RF "On selective achievements" of August 6, 1993 No. 5605-1; Federal Law of the RF "On the ratification of the Eurasian patent convention," of June 1, 1995 No. 85-FZ; Civil Code of the RF. Part I. Federal law of Nov. 30, 1994 No. 5-FZ; Civil code of the RF. Part II. Federal law of Jan. 26, 1996 No. 15-FZ; Tax code of the RF. Federal law of June 13, 1996 No. 63-FZ.

29.   Edict (ukaz) of the President of the RF of May 14, 1998 No. 556 "On the legal protection of the results of R&D work of a military, special, and dual-use nature." Resolution (postanovlenie) of the government of the RF of Sept. 29, 1998, "On first-priority measures for the legal protection of the interests of the government in economic and civil-law circulation of the results of R&D of military, special, and dual-use nature."

30.   "Basic directions of government policy for the economic utilization of the results of S&T activity." Order (Rasporiazhenie) of the Government of the RF of Nov. 30, 2001, No. 1607-r.

31.   *Russian Science at a Glance—2005. Statistical Yearbook,* CSRS, Moscow, 2005, p. 22.

32.   Data supplied by the Fund for Assistance.

33.   Article 66, GK, and also article 24, FZ "Concerning non-commercial organiza-
        tions," No. 7-FZ of Jan. 12, 1996.

6. International Support of Russian Science

1.   George Soros's International Science Foundation.

2.   The International Science and Technology Center, financed by the govern-
        ments of the European Union, Switzerland, the United States, Canada, Japan,
        Norway, Korea, and Russia.

3.   The International Association for the Promotion of Cooperation with Scien-
        tists from the Independent States of the Former Soviet Union, financed by the
        European Union.

4.   A. Kozhevnikov, *Filantropiia Rokfellera i Sovetskaia Nauka* (*Rockefeller phi-
        lanthropy and Soviet science*) (St. Petersburg–Moscow: Mezhdunarodnyi fond
        istorii nauk, 1993).

5.   L. Graham, "How Valuable Are Scientific Exchanges with the Soviet Union?"
        Science 202 (October 27, 1978): 383–390.

6.   G. Sher, "U.S.-Russian Scientific Cooperation in Changing Times," *Problems of
        Post-Communism* 51, no. 4 (July–August, 2004): 25–33.

7.   *Review of U.S.-U.S.S.R. Interacademy Exchanges and Relations* ("The Kaysen
        Report"), (Washington, D.C.: National Academy of Sciences, 1977).

8.   Andrei Sitov, "Russia, USA open New Epoch in Science & Technology Coop-
        eration," ITAR-TASS News Agency, January 13, 2006.

9.   Reorientation of the research capability of the former Soviet Union: A re-
        port to the Assistant to the President for Science and Technology, Results of
        a workshop on March 3, 1992, sponsored by the NAS, NAE, and IOM, Wash-
        ington, D.C.: National Academy Press. For Russian reactions, see *Radikal* no. 17
        (May 1992) and *Nauka v Sibiri* no. 14, no. 17 (April–May, 1993). For a later U.S.
        report on a similar theme, see "Partners on the Frontier: The Future of U.S.-
        Russian Cooperation in Science and Technology" (Washington, D.C.: Office
        for Central Europe and Eurasia, National Research Council, 1998).

10.   *Mezhdunarodnye, regional'nye i natsional'nye organizatsii, fondy i programmy:
        Spravochnoe izdanie* (*International, regional, and national organizations, foun-
        dations, and programs: A reference book*) (Voronezh: VGU, 2002).

11.   For example, about 40 percent of the CRDF budget consists of contract works
        for the U.S. State Department (evaluation of proposals for the ISTC) and also
        the U.S. Department of Defense (logistics, organizational work).

12.   ISF Annual Report—1993, p. 1.

13.   Irina Dezhina, *The International Science Foundation: The Preservation of Basic
        Science in the Former Soviet Union* (New York: ISF, 2000).

14.   A. Alakhverdian, I. Dezhina, A. Iurevich, "Zarubezhnye sponsory rossiiskoi
        nauki: vampiry ili Santa-Klausy?" ("Foreign sponsors of Russian science: Vam-
        pires or Santa Clauses?") *Mirovaia ekonomika i mezhdunarodnye otnosheniia*
        no. 5 (1996): 35–45.

15. I. Dezhina, *The International Science Foundation: The Preservation of Basic Science in the Former Soviet Union* (New York: ISF, 2000).

16. "Poka eshche pytaius' ubedi't" ("I am still trying to convince people"), interview of Zhores Alferov by Tatiana Ernova, *Nevskoe vremia* no. 85 (May 13, 1995): p. 5.

17. On the ISTC see Glenn Schweitzer, *Swords into market shares: technology, economics, and security in the new Russia* (Washington D.C.: John Henry Press, 2000); and his *Experiments in cooperation: Assessing U.S.-Russian programs in science and technology* (New York: Twentieth Century Fund Press, 1997).

18. *International Science and Technology Center: Annual Report for 2005*, p. 26.

19. "The Governing Board of the ISTC," (June 28, 2006) http://www.istc.ru/istc/sc.nsf/news/20060706-1 (accessed November 2007).

20. G. Knezo, "Research and Development Funding in a Constrained Budget Environment: Alternative Support Sources and Streamlined Funding Mechanisms," CRS 96–340 (April 5, 1996), Library of Congress, Washington, D.C., pp. 24–25.

21. Ekspress-informatsiia TsISN, "Monitoring reformirovaniia rossiiskoi nauki," no. 11/12, 1997.

22. L. Romankova, "Problemy vosproizvodstva nauchno-pedagogicheskogo potentsiala vysshei shkoly na sovremennom etape" ("Problems in maintaining the scientific and teaching level in higher education of the moment"), *Podgotovka nauchnykh kadrov v sisteme vysshego obrazovaniia Rossii* (Moscow: INION RAN, 2002), 121.

23. For more information on the 7th Framework Program, see http://cordis.europa.eu/fp7/home_en.html.

24. *Nauka Rossii v tsifrakh—2004 Statisticheskii sbornik* (Moscow: TsISN, 2004); *Nauka Rossii v tsifrakh—2005 Statisticheskii sbornik* (Moscow: TsISN, 2005).

25. Ibid.

26. Statistics are for 2002. *OECD Science, Technology and Industry Outlook*, OECD, 2004, p. 193.

27. Several European organizations, such as DAAD, are also giving more attention to outlying regions of Russia.

28. "Podgotovka k novomu desiatiletiiu program obmena mezhdu SShA i Rossiei" ("Preparation of a program for a new decade of U.S.-Russia exchange"), Materialy simpoziuma, Moscow, 2004, pp. 25, 37.

29. V. I. Bakhmin, *O fondakh v Rossii* (*On foundations in Russia*) (Moscow: Logos, 2004), 47.

7. Strengthening Research in Russian Universities

1. There were, of course, exceptions to this generalization. Good research has been done in Russian universities, and in some fields, such as mathematics, this emphasis on youth was stronger than in others.

2. For further information see: http://www.nsf.gov/od/oia/programs/stc/index.jsp.

3. Statistics for June 2005 from the Russian Ministry of Education and Science.

4.   For a positive evaluation of the BRHE program in a U.S. source, see Bryon McWilliams, "In Russia, a Model Program Reunites Research and Academe," *The Chronicle of Higher Education,* March 16, 2007, A40–A42.

5.   S. Beliaeva, "Stimuly svobody: Amerikanskii fund vydelil novye granty universitetam Rossii" ("Stimuli for freedom: An American foundation gave new grants to Russian universities"), *Poisk* no. 23 (June 10, 2005), 15.

6.   "The Bologna Process from a Norwegian Perspective: Towards a European Higher Education," fact sheet, the Bologna–Bergen Summit, Ministry of Education and Research, 2005. http://www.bologna-bergen2005.no/Docs/Norway/041014Fact_Sheet_Bologna-Process.pdf.

7.   The Russian government claimed that the support of integration was continued in less obvious form through the Federal Goal-Oriented Program titled R&D in Priority Directions of Science and Technology, but even the latter program was ended in 2006.

8. Impact of International Activities on Russian Science

1.   A. Iu. Chepurenko and L. M. Gokhberg, eds., *Vosproizvodstvo nauchnoi elity v Rossii: rol' zarubezhnykh nauchnykh fondov (na primere Fonda im. A. Gumbol'dta)* (*Reproducing the scientific elite in Russia: the role of foreign foundations*) (Moscow: RNISiNP, 2005), 153.

2.   I. Dezhina, "Molodye kadry v rossiiskoi nauke: kak ikh sokhranit'?" ("Young people in Russian science: How to preserve them?"), 128.

3.   U.S. Civilian Research and Development Foundation, http://www.crdf.org/Events/conference2001.html (accessed in 2005; at time of publication, website was not available).

4.   I. Dezhina, *The situation of women researchers in Russian science,* 61–81.

5.   *Nauka v regionakh rossii: statisticheskii sbornik* (Moscow: TsISN, 2004), 112.

6.   I. Dezhina, "Pomoshch' nauke v regionakh: tsel' ili sredstvo?" ("Help for Russian science: Goal or means?"), in M. V. Alfimov and V. D. Novikov, *Granty RFFI: rezul'taty i analiz: sbornik* (Moscow: Ianus-K, 2001), 81.

7.   V. Kostiuk, Report of the chief scientific secretary of the Presidium of the Russian Academy of Sciences, *Poisk* no. 20 (May 20, 2005), 6.

8.   U.S. Civilian Research and Development Foundation; http://www.crdf.org/Evaluation/CGPIIEvaluationFINAL.pdf (accessed in 2007; at time of publication, the website was unavailable).

9.   M. Blokhuis, "Ten Years of Dutch-Russian Scientific Cooperation: An Internal Evaluation of the Dutch-Russian Scientific Cooperation Programme, 1999–2003" (The Hague: NWO, December 2003).

10.   "Assessment of the Contemporary Issues Program: Final Report," Aguirre International, IREX, March 2005, p. 4.

11.   K. Schuch, "Joint RTD Projects Between the EU and Eastern Europe: What Does Really Matter?" in *Dialogue on S&T between the European Union and the Russian Federation* (Moscow, CSRS, 2002), 133–146. http://www.csrs.ru/English/Public/Analit/2002/Dialogue/034KlausSchuch.pdf.

12.   Chepurenko and Gokhberg, eds., *Vosproizvodstvo nauchnoi elity* (*Reproducing the scientific elite*), 93.

13.   "Survey among Dutch and Russian Participants," NWO, The Hague, October 1997; Blokhuis, "Ten Years of Dutch-Russian Scientific Cooperation."

14.   R. Burger, "Deiatel'nost' INTAS v voprosakh innovatsii i valorizatsii: Presentatsiia na seminare Fonda sodeistviia razvitiiu malykh form predpriatii v nauchno-tekhnicheskoi sfere i CRDF" ("The work of INTAS in innovation and valorization"), Nauchnyi park MGU, Moscow, June 6, 2005.

15.   "Assessment of the Contemporary Issues Program: Final Report," iii.

16.   INTAS, http://www.intas.be/documents/news/PR/09-06-05-GA_statement_Ext_Eval.pdf (accessed in 2005; not available at time of publication, as this organization no longer exists).

17.   ISTC Annual Report, 2005, p. 14. http://www.istc.ru/ISTC/sc.nsf/AR-2005-en.pdf (accessed November 2007).

18.   S. I. Roslovaia (specialist on applied technology, ISTC), interview by Irina Dezhina, July 3, 2003.

19.   A. G. Ivanchenko (representative of the public relations department of the ISTC), interview by Irina Dezhina and Loren Graham, September 9, 2003.

20.   Iu. O. Lebedev, "Iuridicheskie osnovaniia vozniknoveniia intellektual'noi sobstvennosti pri issledovaniiakh, vypolniaemykh v kachestve dogovornykh obiazatel'stv—sovremennoe sostoianie v Rossii" ("The legal foundations of intellectual property arising in research carried out on contract bases—the current situation in Russia"), Dokument No. 5, Seminar of the Office of Economic and Statistical Research "Prava intellektual'noi sobstvennostii i gosudarstvennoe finansirovanie issledovanii v Rossii" ("Intellectual property rights and state financing of research in Russia"), Obninsk, October 22–23, 1996, pp. 9–10.

21.   ISTC Annual Report, 2004, http://www.istc.ru/ISTC/sc.nsf/AR-2004-en.pdf (accessed November 2007).

22.   Deborah Y. Ball and Theodore P. Gerber, "Russian Scientists and Rogue States: Can Western Assistance Reduce the Proliferation Threat?" *International Security* 29, no. 4 (Spring 2005): 55–70.

23.   Gerson Sher (founder of CRDF), interview by Irina Dezhina and Loren Graham, February 2005.

24.   Interview of program officers by Irina Dezhina and Loren Graham, Department of International Programs, NSF, August 19, 2004.

25.   Data supplied by Jean-Michel Kantor (one of the leaders of Pro Mathematica).

26.   P. Idenburg, P. Stalnacke, K. Schuch, "External Evaluation Report on the Programme of the INTAS (1993–2003)," INTAS, October 1, 2004, p. 9.

27.   Chepurenko and Gokhberg, eds., *Vosproizvodstvo nauchnoi elity* (*Reproducing the scientific elite*), 92.

28.   http://www.crdf.org/Evaluation/CGPHEvaluationFINALSummaryPage.htm.

29.   I. Dezhina, " 'Utechka umov' " iz postsovetskoi Rossii: evoliutsiia iavleniia i ego otsenok" ("Brain drain from post-Soviet Russia: The evolution of the phenomenon and its evaluations"), *Naukovedenie* no. 3 (2002): 46–47.

30. V. Zinov, "Sluzhba kommertsializatsii nauchno-tekhnicheskikh razrabotok v institutakh Rossiiskoi akademii nauk" ("Commercialization service of R&D in institutes of the Russian Academy of Sciences"), *Kontseptsii* no. 1 (2003): 67.

31. *Fond sodeistviia razvitiiu malykh form predpriiatii v nauchno-tekhnicheskoi sfere: Otchet o deiatel'nosti za 2004 god* (*2004 report of the foundation for small S&T enterprises*) (Moscow: Fond sodeistviia, 2005), 48.

32. Proekt TACIS FINRUS 9804, *Problemy i perspektivy razvitiia rossiiskikh territorii vysokoi kontsentratsii nauchno-tekhnicheskogo potentsiala* (*Problems and projects for development of Russian territories with a high concentration of S&T*), Vol. 6, *Innovatsionnye tsentry i naukogrady* (*Innovation centers and science cities*) (Moscow: Skanrus, 2001), 3–4.

33. See, for example, A. Iurevich, "Neravnoe ravenstvo: rassloenie rossiiskogo nauchnogo soobshchestva" ("Unequal equality: Stratification in the Russian scientific community"), *Naukovedenie* no. 3 (2002): 70–71.

34. "Survey among Dutch and Russian Participants," NWO; Blokhuis, *Ten Years of Dutch-Russian Scientific Cooperation.*

35. Surveys conducted in 2000–2001 and in 2003 gave practically the same results. See, for example, S. Kugel', "Adaptatsiia rossiiskikh uchenykh k izmeniaiushchimsia sotsial'no-ekonomicheskim usloviiam" ("Adaptation of Russian scientists to changing socioeconomic conditions"), *Naukovedenie* no. 1 (2001): 11; A. Iurevich, I. Tsapenko, A. Prikhid'ko, "Skol'ko i kak zarabatyvaiut nashi uchenye?" ("How much income are our scientists earning, and in what way?"), *Naukovedenie* no. 1 (2004): 69.

36. V. I. Bakhmin, *O fondakh v Rossii* (Moscow: Logos, 2004), 58.

37. Conversations of Loren Graham with scientists in Russia, 1996–2004.

38. This phenomenon is under-researched with reference to activities in natural science, but it has been examined with regard to Western aid to Eastern Europe in general. See, for example, Janine R. Wedel, *Collision & Collusion: The Strange Case of Western Aid to Eastern Europe* (New York: Palgrave), 2001.

39. An example is Guy Chazan, "Russia Holds up Foreign Aid Meant for Democracy Work," *The Wall Street Journal: Europe,* April 3, 2006, p. 1.

9. Conclusion

1. V. A. Sadovnichii, ed., *Obrazovanie, kotoroe my mozhem poteriat'* (*The Education Which We May Lose*) (Moscow: Institut komp'iuternykh issledovanii, Moskovskii gosudarstvennyi universitet im. M. V. Lomonosov, 2002).

2. The only period of time that might compete with today in terms of the integration of Russian science with world science would be 1861–1914, but in our opinion the links are more numerous today. Many thousands of cooperative efforts between scientists in Russia and scientists abroad have been made through the ISTC, ISF, INTAS, CRDF, MacArthur Foundation, Carnegie Corporation, NAS, Russian Academy of Sciences, Russian universities, IREX, BRHE, AID, CERN, DAAD, CNRS, DESY, DFG, FSA, NASA, NIH, NWO, NSF, Max-Planck-Gesellschaft, Humboldt Foundation, Royal Society, British Council, Howard Hughes Institute, and other programs, learned societies, foundations, and government agencies, both Russian and foreign.

Index

LOREN GRAHAM, author of *Moscow Stories* (Indiana University Press, 2006), is a well-known historian of science who taught for many years at the Massachusetts Institute of Technology and Harvard University. He is the author of many books on the history of Russian and Soviet science.

IRINA DEZHINA is Leading Research Fellow at the Institute of World Economy and International Relations, Moscow, and the author of many works on post-Soviet politics and society (in Russian).

Printed and bound by CPI Group (UK) Ltd, Croydon, CR0 4YY

13/04/2025

14656549-0001